U0396190

HILLS AND RAVINES IN THE HEART
Chinese Gardens and Landscape Painting

Edited by XU Yinong, GE Ming and GU Kai

胸中丘壑
中国园林与山水画

许亦农 葛明 顾凯 编

Southeast University Press
东南大学出版社 · 南京

for our teachers

献 给 我 们 的 老 师

Preface

This volume owes its inception to a one-day colloquium in conjunction with an exhibition, "Hills and Ravines in the Heart: Chinese Gardens and Landscape Painting," in Caxton House at London South Bank University on 29 November 2017. Sponsored by the Centre for Language Education Cooperation in the context of "China Studies Programme," the colloquium brought together a small group of international scholars to consider the inextricable and complex links between garden making and landscape painting.

In order to reach a wider audience in China and worldwide, we decided to render the book bilingual in Chinese and English. The texts were originally written in one of the two languages and then translated into the other. In consideration of the possible issue of *les belles infidèles* in textual translation, it has become desirable that, for each text, the original version be given pride of place followed by its translation, so as to do justice to each of the essays in terms of their accuracy, text flow, and idiomatic expression. Also ensued from the bilingual production of the volume is the problem of differences in the sequential order of family and given names between Chinese and Western customs for our contributors. To accommodate both Chinese and English readers, we decided to follow the Western order of names for Western contributors and the Chinese order of names for Chiese contributors except in case of combined Chinese-Western names, and to capitalise all family names for the sake of consistency and clarity.

We are grateful to the Centre for Language Education Cooperation for its generous and invaluable support of this project. Our gratitude also goes to London South Bank University for successfully hosting the colloquium and exhibition, and to the School of Architecture at the Southeast University for co-organising the event and contributing to the publication of this book. Containing essays based on three quarters of the original conference papers, it largely reflects the nuances of the lively discussions that took place during the colloquium, thanks to the diversity of participants and disciplinary approaches. Beyond the contributors, we are obliged for unfailing support and help to many individuals. We thank KONG Yiming, SHEN Yi, SUO Jiani for their efficient and creative design of exhibition installation; HAN Yang (Helen) for her meticulous and tireless logistical work for the entire project; LIANG Jie, TANG Jingyin, SHI Wenjuan for their active participation in the work of collecting and selecting research and reference materials, as well as in the transportation of exhibits; WANG Mingliang, HAN Ying, JI Xiang and HU Jing for their enthusiastic contributions to the miscellaneous tasks for the exhibition and colloquium; LIU Xiaodan and YAN Jia for their assiduous work on the book layout design. Finally, we deeply appreciate the professional and pa tient assistance that DAI Li and WEI Xiaoping, editors from the Southeast University Press, kindly lent to us in the process of publication.

XU Yinong, GE Ming, and GU Kai

卷首语

这本书源于 2017 年 11 月 29 日在伦敦南岸大学卡克斯顿学院举行的为期一天的“胸中丘壑——中国园林与山水画”研讨会暨展览。在中外语言交流合作中心“中国研究项目”的赞助下，研讨会得以使为数不多的国际学者聚集在一起，探讨园林与山水画之间密切、复杂的关系。

为了扩大在中国乃至世界的读者范围，我们决定用中英双语出版此书。论文最初都是用这两种语言中的一种撰写，然后翻译成另外一种语言。考虑到文本翻译中可能出现的语意准确和文字优美难得两全的问题，我们采取了每篇论文原文在前、译文在后的方式，以便充分体现每篇论文自身的准确、流畅和习惯表达。另外，本书双语出版也带来了撰稿人姓名在中西方习俗中排序不同的问题。为了兼顾中英读者，我们决定让西方作者名字遵循西方的名前姓后顺序，中国作者名字遵循中国的姓前名后顺序（中西方混合名字除外）；同时，为了保持一致和明晰，所有姓氏全部大写。

我们感谢中外语言交流合作中心为本项目提供慷慨而不可或缺的支持，感谢伦敦南岸大学作为东道主成功举办研讨会和展览，并感谢东南大学建筑学院协作举办该活动，并为本书出版做出贡献。研讨会上四分之三的原始论文都收集在这里，因此很大程度上反映了当时生动讨论的细微之处，这得益于参会者和其学科研究方法上的多样性。除了供稿者之外，许多个人也为这个项目切实付出了各种努力，我们在此一并致谢：孔亦明、沈祎、索佳妮做出高效且富创意的布展设计；韩阳细致、不懈地主管整个项目过程的所有后勤事务；梁洁、唐静寅、史文娟负责整理、筛选研究和参考资料以及展品运输等工作；王明亮、韩颖、纪翔、胡靖在研讨会与展览期间积极、热诚地参与各项工作；刘筱丹、严佳悉心着力于本书的版面设计。最后，我们向东南大学出版社戴丽副社长和魏晓平编辑表示诚挚的谢意，她们在出版过程中给予我们以专业、耐心的协助。

许亦农、葛明、顾凯

Contents
目　　录

HILLS AND RAVINES IN THE HEART >>

Introduction

XU Yinong

It has long become almost common knowledge that at least since the eighth century CE, garden making and landscape painting in China have been intertwined in their respective developments, a distinctive feature on which few modern and contemporary students of Chinese garden culture have ever failed to touch. For this reason, the theme of this volume is as unoriginal as it is sound. And yet it does afford us a good opportunity to bring together observations of and deliberations on Chinese gardens from contributors of differing disciplinary training—studies of Chinese literary history and art history on the one hand, and studies of architecture and architectural history on the other. Such a scholarly confluence itself embodies one of the objectives of the original colloquium in 2017, on which this book is based: to facilitate direct and close dialogues in the field of Chinese gardens between these modern disciplines, whose divergence bears on both profound and subtle differences in research framework, orientation, methodology, and focus.

The cross-disciplinary approach in this particular context has led to a concomitant advantage, which was another intended objective of the colloquium. Early interest in and knowledge of Chinese gardens had been formed in Europe by the latter half of the nineteenth century from accounts of European travellers—Jesuit missionaries, merchants, diplomats, artists, architects, plant hunters, and so on—who visited China.[1] From the early twentieth century onward, studies of Chinese gardens in Europe and America became more scholarly and sophisticated, but have largely remained in the purview of art historians and, to a less degree, of students of literature. In China, by contrast, this field of studies was initiated in the first half of the twentieth century by architects and architectural historians, and gardens' spatial layouts, atmospheric dispositions and physical elements were the main subject areas of their studies. Such a focus has, thanks to the disciplinary nature of architectural schools, continued to this day, even though the horizon of research has gradually been extended over time to social, historical and cultural aspects of gardens in China. Dialogues between disciplines thus simultaneously bring about dialogues between scholarly cultures of China and the West.

Our choice for the colloquium of the broad theme of garden making and landscape painting, coupled with its cross-disciplinary and cross-cultural intentions, has a cost, however. The seven es says contained in this volume are, unsurprisingly, quite discrete in their topical focus and adopted approach; indeed, they are so discrete that the whole volume becomes, *prima facie*, in want of coherence. In this situation, the common thread by which we tie the essays together has to come from the numerous inextricable links between the two artistic and cultural traditions. Suffice it here to mention just a few: they share many terminologies, concepts and criteria; they have parallel or related trajectories of change in certain historical periods; they both convey a strong sense of the medium of existence, by which the relationships between components are negotiated; they similarly build on "conceived" or "created" nature and invite or require participation from the viewer or visitor, thus relying as much on their creation as on their interpretation; and they resonate with each other in the issues of self-expression, of mode of viewing and experiencing, and of intellectual, cultural and literary association. Each of the essays discusses or at least touches on one or more areas of these links.

Both Chinese and Western scholars have generally accepted the special significance of the centuries of disunity (221–589) for the development of Chinese garden culture. It was a period of fundamental changes of world view. Nature and landscape were gradually perceived and enjoyed for their intrinsic value and beauty, rather than solely as a metaphor for human virtues, an emblem of the power and majesty of the imperial domains, or a simple adjunct to accounts of human activities. This change happened not only in literature, art, and gardening, but also in daily life.[2]

Disillusioned with the Confucian view of the social and political order and with the prospects of officialdom, some men of letters turned to deliberate eccentricity of behaviour, while some others became obsessed with "thinking in the mysterious" (*xuansi*) and "pure conversations"(*qingtan*), both disguising courses of action intended to conceal, but at the same to reveal, their true feelings or understandings through unconventional wit. Still some others became attracted to the time-honoured ideals of reclusion, which was accompanied by an increased predilection for country living and an appreciation of nature. Landscaped mountain estates, gardens, and parks flourished from the fourth century. It was certainly more than coincidental that in this pivotal period landscape painting started to develop, and the first extant discourses on landscape painting appeared. Literati critics and commenters of later times on gardens and landscape paintings often alluded to the literary works and anecdotes of this period, drawing justifications or persuasive powers from these sources for their own concepts or interpretations.

There is no shortage of studies of garden development in this transitional period,[3] but Chen Wei's essay, "Living amid the Mountains and Waters of the Six Dynasties," emphatically places Jinling (present-day Nanjing) in the limelight and discusses how landscape and garden culture unfolded in space and time around this political, economic and cultural pivot. Chen Wei sees that under the volatile political conditions of the time, the convergence in Jinling of clans of nobility, families of literati, and landowning monasteries brought about flourished commerce and a new culture of individual thinking and expression, which in turn provided both the economic basis for booming of country estates and the ideological basis for intellectual pursuit of the individuals. This then led for Chen Wei to the burgeoning of gardens, from imperial to private and to monastery, which was facilitated by the fine natural landscapes of the Jiangnan region. Taking the two often-cited clusters of events as geo-historical markers —the disputatious gatherings of the Seven Sages of the Bamboo Grove around 240–249 in Shanyang, modern-day Henan province, and the lustration festival gathering at Lanting in 353 in Kuaiji, modern-day Zhejiang province—Chen Wei argues for the distinctive and crucial role that Jinling had played in the process of garden and landscape transformation during this period and even thereafter.

At the end of her essay, Chen Wei cites the late Tang poet Wei Zhuang's (836–910) "Terraced City Walls," and remarks that what the poem expresses is not only a fine scenery but also a sort of "the realm of ideas and feelings" and of "spiritual wandering." The idea of "recumbent wandering"[4] in the landscape was already present in an anecdote of Zong Bing (375–443), as recorded in the *Book of Song* and repeated in Zhang Yanyuan's (ca. 815–ca. 877) *Record of Famous Painters of All the Dynasties*.[5] It was not until the 11th century, however, that the concept of landscape painting's capacity to inspire multiple modes of enjoyment fully appeared: "there are landscapes one can walk through, landscapes that can be gazed upon, landscapes in which one may ramble, and landscapes in which one may dwell."[6] This is indeed one of the concepts that theorists and critics of both landscape paintings and gardens have continuously relished, and is once again seriously considered by Minna Törmä in her essay, "Landscapes for Travelling, Sightseeing, Wandering or Dwelling." But Törmä's discussion goes far beyond any simple descriptions of these four modes and their applications; it centres instead on Osvald Sirén's (1879–1966) suggestion that "the enjoyment [of Chinese gardens] may be compared with the study of a landscape painting in the form of a long horizontal scroll,"[7] and explores the extent to which this analogy may be taken as tenable.

To demonstrate and argue for the variety of nuances in both gardens and handscroll paintings and for complex relationships between these two forms of art, Törmä takes the reader on a journey through the landscapes of two specific scroll paintings—*Streams and Mountains without End* by an anonymous painter in the first half of the twelfth century and the 1662 *Dream Journey among Rivers and Mountains*, Number 150, by Cheng Zhengkui (1604–1676)—and, as a parallel approach, through the various quarters of Wangshiyuan (Garden of the Master of the Net). Along these journeys Törmä enriches her line of argument by touching upon a few paired concepts, including Chen Congzhou's "viewing in motion" and "viewing in repose," and the fascination with the polarity between the real and the imaginary in the late Ming. But the most critical and instrumental differentiation Törmä makes is between the "internal viewer" and the "external viewer" of landscape paintings, and by extension, between the visitor and designer of gardens. For Törmä, although "[a] garden is viewed from the point of view of the Traveller (internal viewer) only," in the end she insists that "the viewer of a handscroll or the visitor to a garden oscillates between two points of view."

Differences, along with parallels acknowledged, between the art of garden making and that of landscape painting are emphasised likewise at the beginning of Antoine Gournay's "*Jingye*: Crafting Scenes for the Chinese Garden." Also similar to Törmä's essay, Gournay's implicitly critiques the modern approach that examines garden components in separate categories, such as artificial rocks, waters, plants, and buildings. But the similarities hardly go further, as Gournay directs the reader's attention to the fact that the various components of a garden, in generating what he calls scenes of "the spectacle," "far from being simply juxtaposed or mixed randomly, are indissociably linked and organized within a coherent and constructed whole, intended to produce certain effects." By citing many examples, ranging from imperial to private gardens, and spreading from the south to the north, Gournay analyses at length the various ways in which the components of the garden come together for specific and often intended effects. Whereas some of the devices or measures are aligned with those of landscape painting, Gournay reasonably pays more attention to those that are absent from or at variance with them, illustrating at least three aspects in which "the garden spectacle differs from that of painting": first, the garden addresses all physical senses of the viewer rather than merely the visual; secondly, the garden unfolds three dimensionally, allowing and sometimes requiring the viewer to traverse it bodily; and thirdly, the garden is both spatial and highly temporal to the extent that "one never sees the same garden twice," which somehow reminds us of the words of Heraclitus (ca. 540–ca. 480 BCE) : "You cannot step into the same river twice" (Plato, *Cratylus*, 402a).

Those aspects of temporality discussed by Gournay are all "natural" ones; that is, the garden "is constantly changing, by itself." And yet as far as garden history is concerned, two other aspects of change, though outside the purview of Gournay's essay, have to be taken into account. One of them is related to the mortality of the garden. Any garden boasting a long history must inevitably have undergone over time numerous repairs, restorations and even reconstructions, which would certainly bring to it physical modifications, alterations and transformations. The second aspect is sometimes related to, but not dependent on, the first: physical changes in gardens brought about by historical changes in aesthetic criteria and taste. After all, we are concerned with a vital Chinese garden history, not a moribund one.[8] Craig Clunas, for instance, has demonstrated dramatic shifts around the mid-Ming period in the perception and use of gardens from largely agronomical sites to fundamentally aesthetic ones.[9] We also know that approaches to garden-making once again underwent significant changes in the late Ming and further in the early Qing, this time centred on aesthetic orientations and tastes. Gu Kai devotes his essay, "'Knowing the Painting Vein' and 'Being Like in a Ravine'," to the event of reconstruction around 1666–1668 of one particular garden, Jichangyuan (Garden for Lodging One's Expansive Feelings), and anchoring his analyses and discussions in the historical context of this identified shift in garden design, and especially in the design of rockery.

The mid-seventeenth-century reconstruction work for Jichangyuan was carried out by Zhang Shi, a nephew of Zhang Lian (aka Zhang Nanyuan, 1587–ca. 1671). The work set out a landscape layout and disposition of the central part of the garden which is believed to have lasted to the present day. Gu Kai holds that this project instantiates the traits of "rockery design by what is called the Zhang family", in which two interrelated features are salient. One is the predilection for the created landscapes that follow the *huayi,* "intents in painting,"[10] which was derived from their knowledge of the continuous line of development of landscape painting in the Yuan period. The other is the attention paid to the kind of design that would effect a viewing and physical experiencing of the landscape by walking through the garden, analogous to one's imagined roaming around in the landscape painting. On the basis largely of various textual descriptions in the Qing period and partly of existing physical conditions of the present garden, Gu Kai analyses the possible application in Jichangyuan of these two featured principles in tremendous detail, especially around the major pond and the rockery on its west bank for the former, and in the area of the artificial rocky ravines for the latter.

Zhang Lian and the junior garden designers of his family

were active at a time when landscape painting and garden culture continued to be an integral part of elite life. Also continued and perhaps intensified was the practice of taking these two forms of art by individual scholars as a means of self-expression and self-identification. Combining the two to this end, *biehaotu*, which may be translated as "cognomen pictures," "by-name pictures" or "studio-name pictures," became a fairly popular genre in the second half of the Ming, especially among the painters of the Suzhou School. In this sense, gardens and *biehaotu* are much in common, as Alison Hardie states in her essay, "Gardens Real and Imagined: Gardens, Landscape Painting, and *Biehaotu*," both intended as expressions of the garden owners' and *biehaotu*'s recipients' projected personality and ideals. What is the ontological nature of the landscape images in *biehaotu*, then, and how are they related to the actual gardens of their owners? By analysing thirteen judiciously chosen examples, Hardie explores these two principal questions while briefly tackling a few other related issues in passing. In the end, she reaches a definite, albeit tentative, conclusion that the boundary between *biehaotu* and landscape or garden paintings is highly elastic, and thus a sharp distinction cannot be drawn between them. This is hardly surprising, because Hardie has deliberately considered not only *biehaotu* par excellence—paintings that are inarguably titled with *biehao*—but also "pictures having some of the characteristics of *biehaotu*." One may contend that Hardie is stretching the point to some extent, but I believe such an elaboration to be particularly worthwhile. Indeed, as Hardie acutely observes, "we should be very cautious ... about believing that artists in the Ming (at least with the exception of the remarkable Zhang Hong) were even attempting to represent their friends' or patrons' gardens with any degree of optical faithfulness, though ... we can use some paintings—with due care—as guides to what gardens of the time may have looked like as regards style and layout." I myself would prefer to go still a little further and insist that even if certain artists did make such attempts, we should still not take it for granted that the pictorial representations would truthfully depict or reflect the physical appearance of the gardens, simply because their modes of seeing and thinking and their devices of representation are surely quite different from ours. After all, paintings are but the projections of the artistic minds.

The problems of seeing and thinking are elusive, for on

the one hand our mind is constantly entangled in the perceptual snare of the customarily thinkable and unthinkable, and on the other hand our scrutiny of the issue itself can be easily conditioned by the mode of thinking of our own. I have framed my essay, "Hills and Ravines in the Heart: A World without 'Space'," in this line of thinking, attempting to ask a basic question, or perhaps more accurately, to question a basic assumption: is the modern or Western concept of "space" applicable or relevant to pre-modern Chinese gardens and landscape paintings if we try to read them on their own terms? That concept of space is most conspicuously, though not exclusively, manifested by the use of or reference to linear perspective. There is no doubt that linear perspective had been imported from Europe to China from the late sixteenth century onward, and had met with varied degrees of interest first at the imperial court and then among some artistic circles (especially the artists of woodblock prints in Suzhou) and certain individuals. Kristina Kleutghen is right to note that the presence of perspective in popular art of the Qing period has been understudied, while the development of perspective "from an aspect of specifically technical knowledge limited to the court into an artistic technique found across the Qing Empire" overlooked.[11] It is equally important to note, however, that among the cases of making use of linear perspective, it is accepted fully in some works, partially in some others, and ostensibly in still some others. Unfortunately, even more in paucity are precisely studies of the specific ways in which linear perspective is present or apparently present, and the possible motivations with which this "artistic technique" is employed or suggested, in both court and popular paintings. Such studies would otherwise be able to offer us some clues as to how the pre-modern Chinese painters, as well as Chinese scholars in general, perceive both the physical and pictorial worlds. Less ambitious than this, the arguments presented in my essay, developed through some close pictorial and textual readings, is that the concept of an infinite, isotropic and homogeneous space, no matter how pervasive and deeply entrenched in our modern mind, did not exist in China before the importation of European traditions, and therefore that it can be quite irrelevant to discuss what kind of perspectives are employed in pre-modern Chinese paintings and gardens.

I am certainly not questioning the instrumental values of such a modern concept for analysing and interpreting either landscape paintings or gardens, but merely arguing that, to read pre-modern Chinese artists' works in their own right, we should avoid imposing our modern concept of space on them and thus replacing their ways of seeing and thinking, which were products of their time and traditions, with ours. Indeed, we may choose to bring their work within our perceptual or theoretical frame of reference, but the choice has to be a conscious one, so that we do not lose sight of the fact that they are intellectually different from ours. It is precisely these differences that render their achievements ever more suggestive or even evocative for our own thinking and work. Ge Ming's short essay, "A Brief Introduction on Six Ap proaches to Gardens," seemingly out of place in this collection of scholarly essays, nevertheless presents a case in point. He shows how he derives from landscape paintings and, to a less degree, from gardens his ideas that are centred on a series of conceptual relationships between object and void, location and boundary, distance and connexion, intermediary and partition, etc., and how he applies these ideas to a number of his small-scale building projects. Ge Ming's interpretations of the specific paintings may be highly idiosyncratic and the conclusions he draws from them are entirely personal, but the value of this undertaking is not necessarily diminished, for it demonstrates that pre-modern Chinese gardens and landscape paintings are undoubtedly rich sources of inspiration, for the past, the present, and the future.

序

许亦农

（韩阳 译）

人们早已普遍认识到，至少自公元八世纪以来，中国园林和山水画一直交相辉映地发展。这一引人注目的现象无不为现代和当代中国园林文化方面的学者所提及。出于这个原因，本卷的主题虽憾无新意，却稳妥可靠。而这也确实为我们提供了一个很好的机会，让我们将不同学科领域的供稿者——一些来自中国文学史和艺术史的研究，另一些则致力于建筑学和建筑史的探讨——对中国园林的观察和思考汇集在一起。这种学术汇合本身体现了 2017 年我们的初始研讨会的目的之一：促进这些现代学科在中国园林领域中的直接交流和深入对话，而这种对话的活力得益于这些现代学科在研究框架、方向、方法和重点上的深刻而精微的差异。那次研讨会的内容为这本书的出版奠定了基础。

这一特定语境下的跨学科研究方法同时带来了另一个优势，这也是当时研讨会的预期目标之一。直到十九世纪下半叶，通过访问中国的欧洲游客——耶稣会传教士、商人、外交官、艺术家、建筑师、植物搜寻者等等——的记述，早期对中国园林的兴趣和了解已经在欧洲形成。[1] 从二十世纪初开始，欧美对中国园林的研究变得更加学术化，且更加深刻而成熟，但这些研究主要仍保留在艺术史学家的研究范围内，还有一些则为文学研究者所热衷。相比之下，中国这一研究领域的开端来自二十世纪上半叶的建筑师和建筑史学家，园林的空间布局、氛围倾向和物质要素是其研究的主要议题。由于建筑院校本身的学科性质，这种学术侧重一直延续到今天，尽管研究的视野逐渐扩展到中国园林的社会、历史和文化等方面。学科之间的对话同时也因此带来中西方学术文化之间的对话。

然而，由于我们的研讨会选择了园林和山水画这一宽泛的主题，加之其跨学科和跨文化的意图，所以这么做是有代价的。毫不奇怪，本书收录的七篇文章在主题关注和研究方法上相当离散；的确，它们如此离散，以至于整本书乍看起来缺乏连贯性。在这种情况下，我们把这些文章联系在一起的共同线索不得不来自园林和山水画这两种艺术和文化传统之间众多错综复杂的联系。这里只需提几个：它们分享大量术语、概念和评价标准；它们具有某些历史时期的平行或相关的变化轨迹；它们都传达了强烈的存在媒介的意识，通过这种媒介，各组成部分之间的关系得以建立或维持；它们同样建立在“被构想出的”或“被创造出的”自然的基础上，并邀请或要求观者或访客参与，因此它们的意义既来自它们的创造，也依赖对它们的诠释；在自我表达、观赏与体验方式，以及思想、文化、文学上的关联等方面，它们相互共鸣。本书的每篇文章都至少涉及这些联系中的一个方面，甚或多个方面。

中西学者已经普遍承认分裂割据的那几个世纪（221—589年）对中国园林文化发展的特殊意义。这是一个世界观发生根本变化的时期。自然和景观逐渐因其内在价值和瑰丽而被感知和享受，而不仅仅是人类美德的隐喻、帝国领地权力和威严的标志，也不仅仅是人类活动叙述的简单附属物。这种变化不但发生在文学、艺术和园林方面，而且也出现在日常生活中。[2]对儒家礼教治世理想的幻灭和仕途凶险的认识，致使一些文人墨士刻意转向率直任性、怪诞不羁的行为，而另一些人则沉迷于“玄思”和“清谈”，这两个掩饰做法都意在通过非传统的机智幽默来隐藏、但同时也表露自己真实的内心感受或思想认识。再有一些名士被历史悠久的隐居理想所吸引，伴随而来的是对田园生活的偏好和对自然的欣赏。山居庄园、私家园林和皇家苑囿从四世纪开始蓬勃兴盛。绝非巧合的是，在这个关键时期，山水画开始发展，并出现了现存最早的山水画论。后世关于园林和山水画的文人批评家和评论家经常征引这个时期的文学作品和趣闻轶事，为他们自己的概念或诠释从中汲取合理性或说服力。

关于这一过渡时期里园林发展方面的研究并不缺乏，[3] 但陈薇的"生活在六朝山水间"一文则着重聚焦于金陵（今南京），围绕这一政治、经济和文化中心，探讨景观和园林文化如何在时空中展开。陈薇认为，在当时动荡的政治条件下，贵族、士族和寺院地主聚集金陵，带来了繁荣的商业和个人思想与表达的新文化，这反过来既为庄园的兴盛提供了经济基础，也为个人的思想追求提供了意识形态基础。这使陈薇得出园林蓬勃发展的结论——从皇家园林到私人园林再到寺庙园林，江南地区佳丽的自然景观为此提供了便利。以两组经常被引用的事件作为地理历史标志——即 240—249 年左右在山阳县（地处今河南省）的竹林七贤玄思清谈，以及 353 年在会稽郡（今浙江省）的兰亭禊会——陈薇认为，金陵在这一时期乃至此后的园林和景观转型过程中发挥了独特且至关重要的作用。

文末，陈薇引用晚唐诗人韦庄（836—910 年）的"台城"，并评论这首诗表达的不仅是景致，而且还是一种"意境"与"神游"。景观中"卧游"[4] 的想法早已出现在《宋书》记载的宗炳（375—443 年）的一则轶事里，而这则轶事又再次出现在张彦远（约 815—约 877 年）的《历代名画记》中。[5] 但是，直到十一世纪，山水画能够激发出多种欣赏模式的概念才完整出现："山水有可行者，有可望者，有可游者，有可居者。"[6] 这确实是山水画和园林方面的理论家和评论家不断津津乐道的概念之一，而米娜・托玛（Minna Törmä）在她的文章"可行、可望、可游、可居之山水"中则再次认真考虑了这一概念。但是托玛的讨论远远超越了对这四种模式及其应用的描述，而是关注喜龙仁（Osvald Sirén，1879—1966年）的观点“［从研究中国园林中］得到的乐趣可以比拟于研究横向长卷山水画所带来的乐趣，”[7] 并探讨了这一类比在多大程度上或许站得住脚。

为了展示和论证园林和手卷绘画的诸多细微差别，以及这两种艺术形式之间的复杂关系，托玛带领读者穿越两幅特定的卷轴画作以游历其中的景观——十二世纪上半叶无名画家的《溪山无尽》与1662年程正揆（1604—1676年）的《江山卧游图》第150幅——同时作为一种平行研究的方法，又引导读者穿越网师园的各个不同区域。伴随着这些游历，托玛用一些成双对称的概念来丰富其观点，包括陈从周的“动观”和“静观”，以及明末人们所着迷的真实与虚幻这对相辅相成的两极概念。托玛在山水画“内部观者”与“外部观者”上做出关键的、重要的区分，并将其延伸到园林的游客与设计者上。对托玛而言，尽管“园林仅以‘旅行者’（内部观者）的视点被观看”，最后她仍坚持认为“不论手卷的欣赏者还是园林的游览者都在两种视角之间摆动。”

顾乃安（Antoine Gournay）的“景治：中国园林中的造景技艺”的开篇同样承认造园艺术与山水画艺术的平行相似之处，但也强调二者的差别。与托玛的文章相似的另一点是，顾乃安的论述含蓄地批评了现代研究方法，即分别探讨不同类别的园林组件，如假山、水体、植物和建筑。但其相似之处就此止步，顾乃安引导读者将注意力集中到这样一个事实，即园林的众多不同组成部分在生成它所谓的“奇观”时，“并非简单并置或随意混合，而是相互紧密联系并组织在一个连贯一致的、构建起来的整体中，意在创造某些特殊效果。”顾乃安举了大量例子，类别从皇家园林到私家园林，分布从南到北，详尽地分析了园林的组成部分汇集在一起的各种方式，以产生特定且往往有意而为的效果。虽然有些技巧或手法与山水画的技巧或手法一致，但顾乃安不无道理地更加关注那些山水画中没有或与其不一致的技巧或手法，阐释了至少在三个方面“园林奇观不同于绘画奇观”：首先，园林作用于观者的所有感官，而非只是视觉；其次，园林以三维形式展开，允许且有时需要观者亲身游历；第三，园林不仅具有空间性，也具有高度的时间性，以至于“人绝不会两次见到同一座园林”——这让我们不由得记起赫拉克利特（公元前约 540 年—公元前约 480 年）所说的“人不能两次踏入同一条河流”（柏拉图《克拉底鲁篇》，第 402a 条）。

顾乃安讨论的那些时间层面的问题全都是“自然”的；也就是说，园林“自身在持续改变”。然而，就园林史而言，变化的另外两个方面必须予以考虑，尽管这些超出了顾乃安文章的范围。第一方面，随着时间的推移，任何历史悠久的园林必然不可避免地经过无数次的修葺、修复甚至重建，这肯定会给它带来有形的调整、改变和转型。第二方面有时与第一方面有关，但并不依赖于第一方面：由审美标准和品位的历史变化带来园林的形态变化。毕竟，我们关注的是一个充满生机的中国园林史，而不是一个垂死的中国园林史。[8] 例如，柯律格（Craig Clunas）展示了明代中期在园林的感知和使用方面发生的戏剧性转变，从大抵上的农艺场所转向根本上的美学场所。[9] 我们也知道，在明末和清初，造园的态度和方法再次出现令人瞩目的一些变化，这些变化以审美取向和品位为中心。顾凯的文章“‘知夫画脉’与‘如入岩谷’”专注于 1666—1668 年间一座特定园林——寄畅园——重建的事件，将其分析、讨论锁定在园林设计，尤其是假山设计发生转变的历史语境中。

十七世纪中叶寄畅园的改建由张涟（即张南垣，1587—约 1671 年）的侄子张鉽主持，所营造的园林核心部分的山水格局据信一直延续至今。顾凯认为这是“张氏之山”设计特色的实例之一，突显了两个相互关联的特征。一个是造园叠石遵循山水“画意”[10] 的倾向，而画意来自他们对于元代山水“画脉”的认识和领会。另一个特征是注重利于园林游观体验的山水营建，颇与想象中的漫游于山水画之中相似。顾凯主要根据清朝时期的各种文字描述、同时也在某种程度上根据今日所见的园林布局特点，详细分析了这两种设计特征的基本原则在寄畅园中可能得以运用的情况；第一个特征尤其展现在主池塘和主池塘西岸假山周围的格局中，而第二个特征则更多体现在开凿的谷涧地带的处理上。

在张涟及其族中传人活跃的时代，山水画和园林文化仍然是精英生活中不可或缺的一部分；同时持续发展甚或得以强化的现象，是文人雅士将这两种艺术形式作为其自我表达和自我认同的手段。这两种艺术形式以此结合在一起，“别号图”就成为明代后半期的一种颇受青睐的画题，这在吴门画派的画家当中尤其突出。从这种意义上说，园林与“别号图”很相似，正如夏丽森（Alison Hardie）在其“真实的园林与想象的园林：园林、山水画、别号图”中所言，二者都意在表达园主和“别号图”受画人所标榜的个性和理想。然而“别号图”中山水图像的内在性质是什么？这些山水图像又与园主的实际园子有怎样的关系？通过分析十三个精心选择的例子，夏丽森探讨了这两个主要问题，同时也简要述及其他几个相关问题。最后，她得出了一个明确的——尽管是暂时的——结论，即“别号图”与山水画或园林画之间的界限具有很大的弹性，因此在二者间无法做出清楚的区分。这并不奇怪，因为夏丽森不但谨慎地考虑了那些最名副其实的、无可争辩地以“别号”为画作标题的“别号图”，而且也包括了“具有‘别号图’一些特征的画作”。人们或许会争辩说，夏丽森在其论点阐述上有些夸张失实，但我相信这样的阐述特别值得。的确，正如夏丽森敏锐地察觉到，“我们应该非常谨慎……不可轻易相信明代艺术家（至少除了不同寻常的张宏以外）会试图以某种视觉真实来再现他们的朋友或赞助人的园子，尽管……我们可以利用一些绘画——在保持应有的审慎的前提下——来引导我们对当时园林的风格和布局的认识。”我本人更愿意将观点延展得再远一些，强调即便有些艺术家确实做过以某种视觉真实来再现园林的尝试，我们也不能想当然地认为绘画的再现能够真实地描绘或反映这些园林的形貌，原因很简单：他们观看和思维的模式，以及再现的手法与我们的肯定大不相同。毕竟，画作是艺术思想的外在投射。

观看和思维的问题难以捉摸，因为一方面我们的头脑总是纠缠在由习惯所决定的可以想象和不可想象的感性圈套中，另一方面我们对问题本身的探查很容易被我们自己的思维方式所制约。我的“胸中丘壑：一方没有‘空间’的天地”一文就是基于这个思路，试图提出一个基本问题，或者更确切地说，对一个基本假设提出质疑：面对现代之前的中国园林和山水画，如果我们尝试根据其自身的条件和性质去解读它们的话，那么现代或西方的“空间”概念对于它们是否适用或相关？那种空间概念极其显著的特点之一是它可以通过使用或涉及线性透视彰显出来。毫无疑问，从十六世纪末开始，线性透视就从欧洲输入中国，并首先在宫廷，然后在一些艺术群体（特别是苏州木版画艺术家）和某些个人中引起了不同程度的兴趣。李启乐（Kristina Kleutghen）恰当地指出，清代大众艺术中透视的存在没有得到足够的研究，而透视“从局限于宫廷的具体技术知识的一个方面，发展成为出现在整个大清帝国的一种艺术表现手法”这一问题也被忽视了。[11]然而，同样重要的是应注意到，在利用线性透视的作品中，有一些完全接受它，另一些部分接受它，而还有一些则仅表面上接受它。可惜现在恰恰更缺乏研究线性透视存在或看似存在的具体方式，以及这种“艺术手法”在宫廷和流行绘画中运用或暗示的可能动机。这样的研究将会为我们提供一些线索，了解现代之前的中国画家以及中国学者如何感知物质世界和绘画世界。不过我的文章中阐述的论点可没有如此宏大；通过对一些图画和文本的仔细解读，我仅仅强调，无限的、同向同质的空间概念无论在我们现代思想中多么普遍且根深蒂固，在欧洲传统输入之前，这样的空间概念在中国并不存在，因此讨论现代之前的中国绘画和园林中采用了哪些透视的种类则颇不切题。

我当然不是在质疑现代空间概念作为工具在分析和诠释山水画和园林方面的价值，而只是争辩在解读现代之前的中国艺术家的作品本身时，我们应该避免将我们现代的空间概念强加给他们，从而让他们的观看和思维方式被我们的所取代，而他们的观看和思维方式其实是他们那个时代与传统的产物。我们的确可以选择将他们的作品放在我们的感知或理论参照系内，但这必须是一种刻意的选择，这样我们才不会忽视他们的作品在思想和知识层面上与我们的作品不同这一事实。也正是由于这些差异，他们的成就对我们自己的思想和创作更具启发性、甚至更具感染力。葛明的短文“园林六则提要”在本论文集中貌似格格不入，但仍然呈现了一个很说明问题的例子。他展示了他是如何从山水画里、也偶从园林中衍生出他的想法。他的思路集中在实体与空虚、位置与边界、距离与连接、中介与隔断等一系列概念的关系上，以及他如何将这些想法运用到他的一些小型建筑项目中。葛明对具体画作的解读或许甚为特立独行，而他从这些画作中得出的结论也完全是个人看法，但这种工作的价值并不一定会降低，因为它正好表明现代之前的中国园林和山水画无疑是丰富的灵感源泉——无论对于过去、现在，还是未来。

生活在六朝山水间

陈　薇

如果说秦汉将山水作为一种臆想和观念，那么六朝则将山水作为一种环境和审美。生活其间，风生水起，风华流连，风情万种，风轻云淡。“山水”不仅是秀美的物质形态，更是美好生活的所在、自由性情的道场、理想精神的寄托，由此诞生的中国山水文化和山水园林，绵延流长。不止形态和结构，更是姿势和态度，此转折的朝代是六朝。在六朝三百余年间，人们津津乐道的通常是“竹林七贤”和“兰亭禊会”，其实，从北方山阳到南方会稽，以至于在六朝形成对山水的认知、思想、表达的系统，重要的时空跨越和转折是六朝金陵的生活。

1 风生水起金陵地

在中国园林关于山水的讨论中，很少有人特别关注金陵这个地域在历史过程中的转折和决定性作用，其实，它非常重要，用一个词概括十分贴切，曰：风生水起。

金陵是个烟水城市，城市兴衰和水密切相关。这个位于长江南岸的重要城市金陵（明朝曰南京）是南京最早的名字，《金陵图经》上说：“昔楚威王见此有王气，埋金以填之，故曰金陵”[1]；不过金陵在秦始皇统一中国时，被改名为秣陵县，传说秦始皇南巡时，其忧心忡忡的事，就是金陵的王气，因此开秦淮河以泻之，而秦淮河后来又成为南京的母亲河；其实南京还有更大的水作为天然屏障以生王气，这就是长江，孙权称帝前，于建安十六年（211 年）移治秣陵的次年，在金陵邑基础上修筑著名的石头城，就建在江边，石头城倚山为城，以江为池，控江临淮，形势险要，于公元 229 年孙权在秣陵建业称帝，称建业（图 1）；六朝时发展为都城，曰建康，选址东移，位于覆舟山、鸡笼山之南，“东环平冈以为安，西城石头以为重，后带玄武以为险，前拥秦淮以为阻。”[2] 出城正南门，有一条长 5 里的御道直达秦淮河边，而六朝宫城之北便是浩渺的玄武湖，面积应该是现在玄武湖的 4 倍，大约 2000 公顷。可以想见当时的金陵地，东边钟山紫气东来，南边秦淮蜿蜒宽广，西边向北长江奔腾不息，北边玄武湖湖水波澜。后来南唐、明代、民国均在这个地方建都，无不与这特殊的地域通江连河、山水雍容有关。这里防卫坚固又交通便捷，气候温润又气韵灵动，是山水绝佳的宝地。

六朝金陵地作为都城之前，对于自然山水的关注尤其是对水的赞颂，在社会上层已渐成风气，东汉建安时期大量写景文学作品的出现应是表征，这和秦汉时期将山水作为比德、作为消遣、作为仙境不同，是将自然作为一种独立的审美对象的自觉开始。建安十二年（207 年）曹操的《步出夏门行・观沧海》便具有山水诗的完整审美表达：“东临碣石，以观沧海。水何澹澹，山岛竦峙。树木丛生，百草丰茂。秋风萧瑟，洪波涌起。日月之行，若出其中。星汉灿烂，若出其里。幸甚至哉，歌以咏志。”景色描述有层次有远近，意境表达有雄浑有情怀。六朝时期对于金陵的认识，谢朓的永明体《入朝曲》最为著名：“江南佳丽地，金陵帝王州。逶迤带绿水，迢递起朱楼。飞甍夹驰道，垂杨荫御沟。凝笳翼高盖，叠鼓送华辀。献纳云台表，功名良可收。”其中，“绿水”“御沟”和“华辀”表达出金陵是地的烟水特征，而“朱楼”“飞甍”和“云台”则为帝都应运江南佳丽地的创造，可看出水体对于金陵地的特别意义（图 2）。

2 风华流连贵族群

在山水文化兴起中，贵族群体的出现犹如发动机和酵素，催化了风华流连。其渊源之一是魏文帝曹丕在位，确立了魏晋时期选拔人才的方式即九品中正制，后演化为高门氏族的门阀政治。这种社会基础是士与宗族的结合，便产生了中国历史上著名的士族。东汉的情形是：不是士族跟着大姓走，而是大姓跟着士族走，士族的地位很高。但到了东汉中叶以后，逐渐显示出政权在本质上与士大夫阶层的重重矛盾，最终借着士族大姓的辅助而建立起来的政权，还是因为与士大夫阶层失去协调而归于灭亡。而士大夫经过“王莽篡位”时的浩然裂冠毁冕而遁迹于山林[3]、东汉中晚期对应“主荒政谬”的第二次隐逸之后，士大夫阶层找到了一种合适的生活方式，走向雅逸。晋武帝司马炎在位，灭吴，全国统一，风格宽简，爱慕名士，是魏晋风度的推动者；晋成帝司马衍在位以来，清谈之风日益浓厚，是魏晋思想形成时期；到晋穆帝永和年间，东晋政权趋于稳定，名士生活更为悠闲。如此，士大夫集学、事、爵为一身，在社会上具相当地位，他们的思想、意趣和追求，成就了一代艺术新风，如此的魏晋风度——由慷慨悲凉转向率性不羁，渐呈浩荡之势。

而在金陵这块地方，既有北方晋室南渡大族，又有江东父老南土士族，还有东汉以降佛教盛行，从而贵族、士族和寺院地主势力强大。一方面，经商成风，“浮船长江，贾作上下。”[4] 金陵地水路交通方便，除了粮、布（麻布）、绢、鱼、盐、铁器外，纸张、瓷器以及席、蜡、蜜、漆、竹等已成交易大宗，尤其是茶叶已随纸、瓷之后作为一种新兴商品登市，以长江边、秦淮河及其道路节点分布的集市为突出现象；寺院经济也成为突出的现象；而士族的庄园经济发达，摆脱了对君主的人身依附关系。另一方面，“过江诸人，每至美日，辄相邀新亭，藉卉饮宴”[5]，“清风朗月辄思玄度”[6]，东晋的清谈成为一种风尚和贵族的风华。清谈时，或在林下，或在室内，或在园内，身着宽大长袍，手执白玉拂尘，当时都城建康有三大清谈中心[7]（图 3）：王导、王濛、司马昱，往来的宾客及陪客都是重要人物，一般内容以“三玄”（《老子》《庄子》《周易》）为基底，引发对人生和宇宙的思索。在王导家，有著名的东晋初期清谈盛会，宾客为武昌东下的长史殷浩，作陪有桓温、王濛、王述、谢尚；东晋中期由王濛主持清谈，参与的名士有支遁、许询、谢安，谢安被推举为主持人，清谈内容是王濛展卷的《庄子》其中一篇“渔父”，谢安言谈“才峰秀逸”“萧然自得”[8]，众咸称善；东晋晚期在司马昱第的清谈最为激烈，来者有殷浩、孙盛、王濛、谢安四大名士，开始主要是殷浩的正统玄学与儒教维护者孙盛之间的较量，在众人即将败北时，又请来才华横溢的刘惔，从饭前到夜宴激战，是阵容最强的一次。

不管怎样，这种由于贵族经济发展更由于名士雅集金陵的情形，至少带来几方面的改变：出现一些新增的聚落场所，往往在交通便捷的水路附近，环境清新，或者山地使用增强，出现田园地产，庄园经济扩展；在意识形态上，通过清谈重新审视个人的人生方向和导致自我觉醒，孙绰的《秋日》便反映了对玄理的体悟：“萧瑟仲秋月，飂戾风云高。山居感时变，远客兴长谣。疏林积凉风，虚岫结凝霄。湛露洒庭林，密叶辞荣条。抚菌悲先落，攀松羡后凋。垂纶在林野，交情远市朝。淡然古怀心，濠上岂伊遥。”

3 风情万种生活态

如此背景下以及接续的南朝，金陵士族经营庄园、谈天说地、饮酒吟诗、行书抚琴、鉴赏观物、登山泛舟，乃成为常态，审美的园林应运而生。其中包含对于东汉而降的上层社会之山水审美认知的承接、对金陵山水因地制宜的应用、对于别具一格人生理想追求的崇尚、对于生活情调和趣味的重视，是风情万种的，非生活态而不能够描绘。

皇家园林方面，以华林园为例，从孙吴时期作为吴宫苑始，到东晋简文帝司马昱经常与群臣在华林园赏“翳然林水”，觉“鸟兽虫鱼”[9]，再到刘宋时期大规模扩建：山有景阳山、武壮山、景阳东岭；水有天渊池、花萼池；台有芳香台、日观台、一柱台；楼殿建筑有景阳楼、连玉堂、灵曜殿、光华殿、射棚、凤光殿、礼泉堂、层楼观、兴光殿等[10]，具有多种功能的活动可能。从南朝至宋的何尚之《华林清暑殿赋》描述可以想见园景和夏天活动："逞绵亘之虹梁，列雕刻之华榱，网户翠钱，青轩丹墀，若乃奥室曲房，深沉冥密，始如易修，终然难悉，动微物而风生，践椒涂而芳质，觞遇成宴，暂游累日，却倚危石，前临濬谷，始终萧森，激清引浊，涌泉灌于基扈，远风生于楹曲，暑虽殷而不炎，气方清而含育，哀鹄唳暮，悲猿啼晓，灵芝被崖，仙华覆沼。"[11]萧齐时期，六朝皇家园林大兴，在齐武帝时，社会安定，经济持续发展，仓廪丰实，修造了娄湖苑、新林苑、芳林苑等，其中芳林苑原为齐高帝萧道成旧宅，齐武帝将其改造成重要的皇家园林，禊宴朝臣是重要活动，王融留有《三月三日曲水诗序》，而临水行祭以祓除不详，谓之"修禊"，其始于三国,但此时禊宴内涵远远超过原义而升腾为雅致的文化行为，"禊饮之日在兹，风舞之情咸荡。去肃表乎时训，行庆动于天嘱。载怀平圃，乃卷芳林。"[12]六朝金陵地，除上述皇家园林，还建有西苑、乐游苑、上林苑、玄圃、东田小苑、桂林苑、宫苑、新林苑、博望苑、青林苑、江潭苑、方山苑、白水苑、灵邱苑、建兴苑等20余处。择址主要在玄武湖、清溪等水体附近和钟山、覆舟山、清凉山等山体周围(图4)，将山水之美和园林的花树之美、动物之灵，连为一体，观日出、避暑热、行流水宴、赏自然，是山水园林滥觞和审美的重要转折时期。

私家园林方面，自永嘉南渡到东晋建国，北方大部分名士迁移江南，以金陵为中心的江南秀丽风光成为审美对象，以及对于清谈的推崇，使得私家园林在金陵形成建造高峰，多选择郊野风景佳地建园，典型如谢安，不仅在会稽、东山寓居，在秦淮河边及土山(今江苏江宁)也营建别墅楼馆，"于土山营墅，楼馆林竹甚盛，每携中外子侄往来游集，肴馔亦屡费百金，世颇以此讥焉，而安殊不以屑意"。[13]在行为方式上，多以登山涉水为常态，所谓"谢公屐"为一种适用游山玩水的鞋子，十分普及（图5）。不过在心态上，东晋名士潇洒高傲，玄风飘逸，而南朝时期隐士文化地位提高，备受朝廷重视，《南史》中隐逸比儒林人物多，是为明证[14]，例如梁武帝萧衍对隐士陶弘景极为重视，咨询朝廷大事要到陶氏隐居之所，这也促成至南朝时私家园林勃兴。南朝时期，建康一带私家园林有案可稽的有30余处，可考的园址主要在秦淮河侧、都城外围及山水附近，包括东府城、乌衣巷、土山、钟山、清溪、秣陵、南涧寺、鸡笼山、娄湖、新林、南冈、东田、东篱门等地。在建园风格上，东晋时期大量种植果树的情形，到南朝也改变为注重经营了。如东晋王导建于钟山的西园，“园中果木成林，又有鸟兽麋鹿”[15]；而南朝常用“修理”“穿筑”和“修营”，如《南齐书·豫章文献王传》“北宅旧有园田之美，乃盛修理之”；《陈书·孙瑒传》“庭院穿筑，极林泉之致”；《宋书·谢灵运传》“遂移藉会稽，修营别业，傍山带江，尽幽居之美”[16]。杜牧《润州二首》之一的“大抵南朝皆旷达，可怜东晋最风流”，也似乎是由东晋而南朝的私家园林风格变化的写照。

寺庙园林方面，最兴盛的是萧梁时期。晚唐杜牧描述的“南朝四百八十寺，多少楼台烟雨中”，应是十分形象的金陵是时的寺庙与山水融为一体的情形。寺庙园林或曰山水环境中的寺庙，是金陵六朝寺庙的突出形态，也可以说是景观建筑出现的集中代表。其形成，有几方面的原因：一是皇家园林和私家园舍都有被改造和利用而作为寺庙的风气。如齐梁古刹栖霞寺，为南齐永明七年（489年）为隐士明僧绍舍宅为寺，当时名“栖霞精舍”，该处层峦拱翠，万壑堆云，自成风景，而自永明二年（484年）至梁天监十年（511年）开窟造像，规模初具，更添神采（图6-图7）；光宅寺始建于南朝梁天监七年（508年），原为梁武帝萧衍的故宅，该地前为赤石矶，后为白鹭洲，风景绝佳；又如鸡鸣寺[明洪武二十年（1387年）重建并是名]最初的建造，是西晋年间利用三国东吴的后苑而为，到梁武帝大通元年（527年）就此创建了著名的同泰寺，距离台城极近，方便武帝修行活动（图8-图9）。所以这种寺庙园林仅是建筑性质改变而已，风景依旧。二是寺庙临近六朝金陵地人群相对集中的聚落而建，而生活聚落当时多选择在江河通达、风景宜人之地（图10）。如瓦官寺附近，东吴时期便是吴国重臣张昭、陆机等人的住宅，东晋时丞相王导在该处置官，主管陶器作坊，晋哀帝兴宁二年（364 年）将陶地赐给僧人慧力建寺，此地本为冈地，至南朝时树木繁盛，翔集孔雀形祥鸟而名凤台山，梁武帝时在凤台山上建瓦官阁而成名胜。李白的《凤凰台》脍炙人口："凤凰台上凤凰游，凤去台空江自流。吴宫花草埋幽径，晋代衣冠成古丘。三山半落青天外，二水中分白鹭洲。总为浮云能蔽日，长安不见使人愁。"依稀可见周边的环境与建筑的关系。三是在思想体系上，六朝时期玄学和佛教及道教有相通之处，在人生探求上相互补足，在生活方式上并不抵牾，尤其在山水审美上高度一致，从而六朝金陵的寺庙园林甚至提升和发展了皇家园林和私家园林在人工结合自然方面的水平与成就。

从时间上讲，六朝三百多年间的金陵，皇家园林在东晋时已十分丰富，而私家园林在由晋而南朝的过程中逐步发展，寺庙园林则在南朝晚期迅速发达。

4 风轻云淡山水情

论及六朝山水与审美、玄学与人物、园林与生活，通常评价与关注度最高的是"竹林七贤"和"兰亭禊会"（图 11–图12），它们几乎代表着魏晋风度的高峰。前者，"山水有清音，何必丝与竹。"[17] 七贤是当时玄学的代表人物，放任不羁，以自然山水而怡情，但政治遭遇多为惨烈；后者，王羲之《兰亭集序》有明确表述："此地有崇山峻岭，茂林修竹，又有清流激湍，映带左右，引以为流觞曲水，列坐其次。虽无丝竹管弦之盛，一觞一咏，亦足以畅叙幽情。"（图 13–图 14）一派优雅，一地清流。两者间隔时间一百年左右，地点均为远离城市的自然佳境，一为金陵的北方，另一为金陵的南方。似乎从未有人特别关注过：从"竹林七贤"到"兰亭禊会"，金陵发挥了什么作用？如果说，这两者的生活方式和择址模式代表着一种独特的文人选择，它们的影响和金陵有何干系？

从小的时空上讲，金陵作为六朝政治和经济、文化的中心，将园林的中心进行了转换，分为两个时间段：第一段便是从“竹林七贤”（三国魏正始年间，240—249 年）到“兰亭禊会”（东晋永和九年，353 年），这一百年间，金陵对于自然秉持了特殊意义的审美态度之继承和发扬，并作为清谈中心成为玄学的汇聚地，即本文前两部分论及的对自然的认知和思想体系的形成，这是山水园林的核心，关乎理想的品格、理想的抱负、理想的生活、理想的情感；第二段从兰亭文人盛会到南朝结束又两百年左右，山水意识和山水园林经金陵为中心的发酵，在江南曼然生长，除了本文第三部分表达的金陵园林系统外，广陵（扬州）、吴兴（湖州）、京口（镇江）、吴郡（苏州）等地均有园林建设和不俗表现。江南山水园林便是在六朝经过内涵的深刻转变和园林类型的极大丰富以及人工结合自然的普及化过程，形成的一种独特的文化体系和基本形态的（图 15），从而奠定了江南园林在晚期发达的重要基础。

从大的时空来看，从汉代到隋唐，政治中心从北方长安和洛阳经由江南金陵复归北方长安，六朝建康是不能逾越的时间与空间的聚焦点，中原传统精华在新的地理空间和政治空间及文化空间下得以延续和发展，其中很重要一项，便是六朝文化中“风轻云淡山水情”流传下来。仅曲水流觞而言，从最早的流杯记载“魏明帝时（227—239 年）于天渊池南设流杯沟，燕群臣”[18]，到东晋第三代皇帝成帝咸和四年（329 年）春正月，台城（宫殿）被攻，太极东堂秘阁被火烧尽，二月以建平园为宫，五年（330 年）秋建新宫，于钟川（钟山流下的水）立流杯曲水，宴请众官，再到晋永和九年（353 年）三月三日王羲之集文人于会稽山阴兰亭曲水流觞，蔚然成风，随后至隋大业，“十二年（616 年）春正月，又敕毗陵郡通守路道德集十郡兵近数万人，于郡东南置宫苑，周十二里，其中有离宫十六所，其流觞曲水，别有凉殿四所，环以清流。”[19] 其时空之重要转换便是六朝金陵，汉魏的山水文化种子，因缘历史的风云际会，在这里生根发芽、移花接木、茁壮成长。接续六朝的隋唐文人园、宫苑、私园和风景园，都绕不开金陵（图 16）。除此之外，南朝的绝唱看似戛然而止，而唐代诗人对金陵山水及其园林的追寻和回味、欣赏和感怀，无不反证出六朝曾经的绚烂、内涵的风劲、绵延的魅力、独特的审美、园景的气质，韦庄的《台城》“江雨霏霏江草齐，六朝如梦鸟空啼。无情最是台城柳，依旧烟笼十里堤”，不仅说的是景致，更是一种意境与神游（图 17）。

“生活在六朝山水间”，是六朝园林的本源，是山水园林的诠释，是形态发轫的根本，是金陵的独特贡献，是中国山水文化中永远的画面。

Living amid the Mountains and Waters of the Six Dynasties

CHEN Wei

(translation by ZOU Chunshen)

Landscape was deemed as an imaginative or philosophical thinking during the Qin and Han Dynasties. Progressively, it was viewed as an environment, and was appreciated from an aes–thetic perspective during the Six Dynasties. People who lived in the Six Dynasties had witnessed the surging and booming devel–op- ment of diversified landscape gardens that embodied a time, a lifestyle, a passion, a sober thought and a secluded taste of life. Landscape garden was no longer a material form that manifested the beauty, but a material showcase of a good life, a dojo that gave one's temperament a free go, and a resting place for one's ideal or soul. They eventually forged a Chinese landscape culture that evolved and strengthened during the long course of history. They overpassed the limit of the pure form and structure, and became a posture and an attitude during the Six Dynasties that marked a turning point of Chinese landscape culture. The Six Dynasties survived the ups and downs of some three hundred years, enjoying an array of popular anecdotes like "the Seven Sages of the Bamboo Grove" and "Orchid Pavilion Gathering". It is a fact that people developed their systematic knowledge, thinking and expression of landscape over an extended time and space from Shanyang in the north to Guiji in the south, and further to the Six Dynasties. However, it is the life in Jinling under the Six Dynasties that materialized a major overpass of time and space, and heralded the arrival of a turning point of Chinese landscape culture.

1 Jinling and emerging Chinese landscape culture

In the discussion of Chinese landscape gardens, few people have paid special attention to the pivotal role played by Jinling in becoming a turning point of a historical process. The issue is substantively relevant, as Jinling used to be the place that heralded the booming development of Chinese landscape gardens.

Jinling was an ancient city famed for the scenes of curling cooking smoke rising over the vast water surface. Water was closely associated with the fate of the city. As an important city physically sitting on the southern bank of the Yangtze River, Jinling was the earliest name of Nanjing before it was called so in the Ming Dynasty. According to the *Jinling Atlas*, "In the old days, King Wei of the Chu State believed the place had an aura of kingdom. He buried gold in the place, and named it Jinling"[1]. It was later renamed as a Moling County under the unified reign of Emperor Qinshihuang. It was rumored that Jinling's kingdom aura made Emperor Qinshihuang worried when he visited the city in his southern tour. He ordered to dig out a Qinhuai River to dilute the aura. The River later became the maternal river of Nanjing. In fact, Nanjing was backed by an even larger water body that served as a natural barrier and that made it enjoy the aura of a kingdom. The water was the Yangtze River. Before claiming himself an emperor, Sun Quan built a stone city on the premises of former Jinling County in 212 CE. The famous stone town sat on the riverside with mountains as the backing and rivers as the moat, enjoying a strategic position that controlled two major rivers (Figure 1). Sun Quan announced himself emperor in the Moling Township in 229 CE. The town was later renamed Jianye. It became a capital city called Jiankang during the Six Dynasties. The city stretched eastward to the south of mounts. Fuzhou and Jilong. "The city was built surrounding the eastern flat hills for a comfortable stay, backed by imposing rocky mountains in the west. It had the buffer zone protection of Xuanwu Lake in the rear, and the blocking capability of the Qinghuai River in the front."[2] Out of the front southern city gate was a 5–li long royal path leading to the Qinhuai River. In the north of the royal palaces sat Xuanwu Lake that possessed a vast water surface around 2,000 hectares that was four times the water surface of today's Xuanwu Lake. One can imagine the kingdom aura prevailing across Jinling at the time: a purple aura stemming from the Mount. Zhong in the east, the vastness extended from the meandering Qinhuai River in the south, the roaring northbound Yangtze River in the west, and the rippling Xuanwu Lake in the north. The city continued to serve as a capital city during the Southern Tang Dynasty, the Ming Dynasty and the Republic of China thanks to its geographic connections to rivers and mountains. It was a well-protected place enjoying easy access and warm/dynamic climate. It was a vast piece of treasure land in terms of landscape.

Even before the Six Dynasties making it a capital city, Jinling was famed for the trendy tastes that the upper class had steadily nurtured for cherishing the beauty of natural landscape, in particular, the worship of water. Numerous literary works stemmed from the Jian'an period of the Eastern Han Dynasty had detailed landscape descriptions. Unlike the literary works of the Qin and Han Dynasties that borrowed landscape to suggest virtues, pastime, or an imaginative fairyland, they started to view and value natural landscape as an independent aesthetic object. For example, a perfect aesthetic expression of landscape was made in *Summer Outdoors: Vast Ocean View* written by Cao Cao (the founder of the Three Kingdoms) in 207 CE. It stated that "One can have a bird's view of the vast ocean surface over the Jie Shi in the east. Water flows here and there dotted with hills and islets. Woods grow prosperously with the affluent presence of vegetations. One watches ocean ripples and waves amid crisp autumn winds. The sun, the moon, the stars and even the galaxy are seemingly part of them. I feel fortunate to see the scenes, and would like to praise them through poetic lines". Cao's description of the scenes tells the range of distance, and his expression of mood betrays his ambition and passion. Xie Tiao's Yongming style "*Song of Dynasty*" had the most famed description of landscape in Jinling during the Six Dynasties. It stated: "The Southern Yangtze River Delta region is a rich and beautiful place. It used to be the main capital of numerous Dynasties. The meandering river flows with green mossy water, and red tiled buildings loom up amid high mountains. The royal path passes through the impressive flying roofs, with willows shading the moat that flows throughout the palace garden. The soothing bamboo pipe sound, joined by light but dense drum sound, sent my gorgeous boat away. I stood in the royal court, and my offer of advice was adopted. I harvested both fame and fortune." Of the lines, the green water, moat and gorgeous boat suggests the water signature of Jinling, while the red tiled buildings, flying roofs and royal court glorifies the royal creations adapted to the beautiful local landscape in the Southern Yangtze River Delta region. Undoubtedly, water has a special meaning to Jinling (Figure 2).

2 Nobles and landscape gardens

In the rise of landscape culture, the emerging aristocratic class added both impetus and enzyme to catalyzing the development of landscape gardens. Under the reign of Emperor Wendi (Cao Pi) of Wei Dynasty, a nine-level official grading regime was created to select talented people. It was a contribution to the subsequent establishment of the so-called aristocratic politics represented by height class family. It laid groundwork for literati and clans working together, and generated the mixed literati-officialdom clans in the Chinese history. During the Eastern Han Dynasty, big-name families did not take the lead followed by the nobles, but rather on the contrary, thanks to the raised social status of nobility. Unfortunately, things changed in the mid Eastern Han Dynasty, where ruling authorities worried about the increased substantive discrepancies with the literati-officialdom class. Consequentially, the regime that was established by taking advantage of the strength of literati-officialdom class and bigname clans perished, due to lacking of effective coordination between the literati-officialdom class and the regime. Meanwhile, the nobles had no other choice but to escape to the mountains after the repeated failure of power games.[3] During the mid-late Eastern Han Dynasty, after the failed attempt to advocate the so-called famine politics, the literati-officialdom class resorted to a free and unfettered way of life that matched their status at the time. Emperor Wudi (Sima Yan) of Jin Dynasty eliminated the State of Wu, and realized the unification of the whole country during the Jin Dynasty. Emperor Wudi advocated simplicity and lenience, and cherished the nobles. He was a promoter of the Wei-Jin Style. During the time of Emperor Chengdi of Jin Dynasty, idle talk became increasingly popular, and led to the formation of Wei-Jin Doctrine. The Eastern Jin Dynasty became relatively stable under the reign of Emperor Mudi. The nobles lived a leisured life, happy with their raised social status thanks to the fact that they embodied the combination of literati, officialdom and noble rankings. Their philosophy, interest and pursuit created a novel artistic aura. The Wei-Jin Style that had changed its connotations from generosity/sadness to straightforward/unfettered became more popular among the literati-officialdom class.

Jinling used to be a rendezvous entertaining both the royal families moving from the north to the southeast and the bigame clans migrated from the eastern side of the Yangtze River. The nobles, literati and monastery landlords were mixed up and became a powerful influence. On the one hand, doing business was very popular at the time. "The Yangtze River enjoyed the busy flow of commercial boats."[4] The easy waterway access made Jinling a busy port for trunk commodity transactions that covered numerous items, including grain, cloth (linen), silk, fish, salt, ironware, paper, porcelain, straw mat, wax, honey, bamboo, among others. In addition, tea became a new popular commodity on the market, followed by paper and porcelain. The open markets along the Yangtze River, the Qinhuai River and major traffic conjunctions became increasingly busy and popular. Meanwhile, the monastery economy stood out as a phenomenon. The plantation economy created by the literati-officialdom class freed the owners from their physical dependence on the monarchs. On the other hand, "Crossing the river, the nobles would invite each other to drink at the new pavilion on sunny days. They would sit on the grass, drinking and eating." [5] And "During a night with bright stars and breezes, I would miss my old friends." [6] Idle talk became a fashion and an aristocratic elegance during the Eastern Jin Dynasty. Idle talk would be made in the woods, or indoors, or in the garden. The nobles would dress up in large robes, holding white jade whisks in hand. At the time, Jinling had three major rendezvous for idle talk at Wang Dao, Wang Meng and Sima Yu's residences. Both the speakers and audiences were important Figures They would speak about their thoughts on life and the universe based on three classic books[7] (Figure 3): *Laozi, Chuang-tzu, and the Book of Changes.* In the early Eastern Jin Dynasty, Wang Dao hosted an idle talk gathering at his residence. The talk was led by Yin Hao, a former high ranking government official, with the participation of other celebrities, including Huan Wen, Wang Meng, Wang Shu, Xie Shang, among others. During the mid Eastern Jin Dynasty, Wang Meng sponsored an idle talk gathering at his residence joined by celebrities, including Zhi Dun, Xu Xun, Xie An, among others. The talk was chaired by Xie An, discussing an article named "Fishing Father"in the Book of *Chuang-tzu*. Xie's articulate and unfettered talk of the article won the participants' applause. In the late Eastern Jin Dynasty, Sima Yu had a hot idle talk gathering at his residence, with the participation of four celebrities, including Yin Hao, Sun Sheng, Wang Meng and Xie An. The debate was made mainly between the orthodox metaphysics represented by Yin Hao and the Confucianism led by Sun Sheng. On the brink of losing the argument, the talented and resourceful Liu Dan was invited to join the debate that lasted till the dinner time. It was the idle talk attended by the strongest lineup[8] in the city, unveiled a conflict between metaphysics and Confucianism.

The nobles' gatherings, a by-product of aristocratic economy, brought up some changes. More gathering venues were added, mostly having an easy waterway access. The enhanced use of mountains or hillsides led to the booming development of pastoral real estate properties, which in turn spurred up the plantation economy. Ideologically, idle talk allowed the participants a new perspective of life and an experience of self-awakening.

3 Living amid landscape

During the following Southern Dynasties, a range of phenomena became a new normal in the city, including literati-officialdom's plantations and their leisured life styles featured with idle talk, poetry making/reading over drinks, calligraphy writing, musical instruments playing, exotic things appreciation, climbing, boating, among others. Progressively, people started to build the landscape gardens with an aesthetic taste. It continued the upper society's practices that started from the Eastern Han Dynasty, including aesthetic perception of landscape, tailored local applications of landscape in Jinling, spiritual pursuit of unique lifestyles/ideals and enhanced attention to the aura and taste of life. They stemmed from life, and were colored/explained by life.

Among the royal gardens built at the time, the Hualin Garden deserves being an example. Started from the time of Sun Wu as a Wu Palace Garden, the Garden used to serve as a venue where Emperor Jianwendi and his ministers had a sightseeing tour amid the lush woods and birds/animals during the Eastern Jin

Dynasty.[9] Having had a facelift and expansion during the Liu-Song Period, the Garden was aligned with three large rockeries, including Jingyang, Wuzhuang and Eastern Jingyang Rockeries, and some water scenes, such as Tianyuan Pool and Flower Pool. It was also built with an array of structures, including Fragrance Platform, Sun Observing Platform and Pillar Terrace, in addition to myriads of functional halls or pavilions, including Jingyang Building, Lianyu Hall, Lingyao Hall, Guanghua Hall, Shooting Room, Fengguang Hall, Liquan Hall, Sight-viewing Pavilion, Xingguang Hall, among others.[10] One can imagine the scenes and summer activities in the Garden based on "An Essay about the Hualin Garden" by He Shangzhi, a government minister during the Liu-Song period, where it stated that "The structures are built with rainbow-like beams, exquisite carved rafters and carved windows with lotus patterns, in addition to the blue short walls and red stone steps. The rooms inside are built deep under a nice planned design. Seemingly easy to build, it is difficult for a layman to see the whole picture. A gentle move brings up the breeze that carries the fragrant smell of the walls. People clicked their wine cups at the feast during their short stay. The site faces a deep valley while backed by huge rockeries surrounded by lush vegetation. The gushing spring water flows in front of the structures, and the winds coming afar blow in through the columns, making a hot summer not deeply felt. The site enjoys fresh and comfortable aura, where one can hear birds singing in the evening and apes crying in early morning. The cliffs are covered by densely growing glossy ganoderma, with flowers seen blossoming in the pond."[11] Under the reign of Emperor Wudi of Qi Dynasty, the country enjoyed a sustained social stability and economic prosperity that guaranteed a good life. The Xiao-Qi period marked the flourishing of royal gardens in the Six Dynasties, with numerous gardens being erected, including the Louhu Gar–den, Xinlin Garden, Fanglin Garden, among others. Of them, the Fanglin Garden was built on the premises of Emperor Gaodi's old residence. Emperor Wudi turned it into a functional royal garden for feasting the ministers of the royal court. Wang Rong, a writer at the time, wrote a preface to the collection of poems on "The Water Entertainment Festival Held Regularly on March 3rd". The Festival was originally created in the period of Three Kingdoms to let folks entertain themselves along the riverside, in an attempt to drive out the evils. However, the feast arranged for the occasion had a connotation far exceeding the original context, and became an elegant cultural event."Folks got together drinking and dancing out the evils. They acted seriously to ask for the blessings of Heaven. It was an impressive visit, and I wrote this to depict the Fanglin Garden."[12] During the Six Dynasties, 20 more gardens were built in the Jinling area in addition to the above-mentioned royal gardens, including the Xi Garden, Leyou Garden, Shanglin Garden, Xuan Garden, Petite Dongtian Garden, Guilin Garden, Palace Garden, Xinlin Garden, Bowang Garden, Qinglin Garden, Jiangtan Garden, Fangshan Garden, Baishui Garden, Lingqiu Garden and Jianxin Garden, among others. The gardens were mainly built along the waters, such as Xuanwu Lake and the Qing Stream, or surrounding the hills or mountains, including Mount. Zhong, Mount. Fuzhou and Mount. Qingliang (Figure 4). They were built to embody the combined beauty of natural landscape and in-garden flowers and trees, and animal spirit. They enabled folks to watch sunrise, sheltering from heat, having a dinner by the waterside and enjoying the beauty of nature. It made an important turning point that people started to view and use gardens in an aesthetic manner.

As far as private gardens are concerned, the inflow of literati-officialdom (from the north to the Southern Yangtze River Delta region during the massive southbound migration to the founding of the country in the Eastern Jin Dynasty) allowed more aesthetic exposure of the beautiful landscape along the Southern Yangtze River Delta region centered around Jinling, and unfettered freedom of idle talk. The development, as such, spurred up the creation of more private gardens in Jinling. Most private gardens were built in the suburban wildness dotted with beautiful landscape. A typical case is Xie An's gardens. He not only built private gardens on the premises of his residences in Guiji and Dongshan, but also built villas and temples on the bank of Qinghuai River or on an Earth Hill (in today's Jiangning, Jiangsu). He "built villas on an Earth Hill, dotted with buildings and pavilions shaded by woods and bamboos. He invited friends and relatives to visit the villas, where he feasted them with the delicacies that would cost him dearly. Folks ridiculed him about it, though ignored by An."[13] At the time, hill climbing and water wading were the main sports at private gardens. For example, "Xie's Shoes" is an invention made to accommodate the needs of touring around the wildness (Figure 5). In the context of mentality, the Eastern Jin celebrities liked to play unfettered, arrogant, mysterious and elegant, while the Southern Dynasties hermits enjoyed a raised cultural status and enhanced royal attention. In the *History of South*, the fact that more hermits were mentioned than literati makes a proof.[14] For example, Emperor Wudi of Liang Dynasty paid great attention to the advice offered by a hermit named Tao Hongjing. The Emperor would consult the hermit about the court events at his secluded residence, which spurred up the booming of private gardens during the Southern Dynasties. During the period, at least 30 private gardens were built across the Jiankang area. The ruins of those gardens were mainly found along the banks of the Qinhuai River, or in the suburbs, or nearing natural waters or mountains, including the Dongfu Urban Section, Wuyi Lane, Earth Hill, Mount. Zhong, Qing Stream, Moling, Nanjian Temple, Mount. Jilong Hill, Lou Lake, Xinlin, Nangang, Dongtian, Dongli Gate, among others. Massive fruit trees growing was a common practice during the Eastern Jin Dynasty, though the focus was shifted to business operations during the Southern Dynasties. For example, Wang Dao, an Eastern Jin politician, built a West Garden near Mount. Zhong, where he "grew massive fruit trees and fed birds/animals, including elks."[15] During the Southern Dynasties, the terms, such as "trimming" "pond digging" "rockery setting" and "camp building", were commonly used. For example, in a book telling the story about Yuzhang Wenxianwang, it said, "The northern residence had a beautiful garden because of frequent trimming." In another book about Sun Chang, it stated, "Folks dug the pond, set up the rockeries and introduced the spring water into the garden." In a book about Xie Linyun, it revealed that "He moved to Guiji, where he built a villa along the river and against the mountain, and made himself a secluded shelter." [16] In one of two Du Mu's poems on Runzhou, it claimed that"Most Southern Dynasties folks are open-minded, while Eastern Jin folks tend to be graceful," suggesting the style differences of private gardens during the Eastern Jin and Southern Dynasties.

Now come the gardens built by temples. The Xiao-Liang period marked the most prosperous development of temple gardens. In the late Tang Dynasty, Du Mu depicted in a poem that "The Southern Dynasties had four hundred and eighty temples, and how many halls and platforms standing amid misty rains" which presented a vivid combined landscape image of the temples amid mountains and waters in Jinling. Temple gardens, or the temples sitting in a theater surrounding by mountains or waters, created a prominent temple form in Jinling during the Six Dynasties. It stood out as landscape structures. Its formation can be attributed to the following facts: first, both royal gardens and private residences could be transformed into or used as a temple. For example, hermit Ming Sengshao donated his residence to be a temple in 489 CE, with a name of Qixia Residence. It was later renamed as Qixia Temple. Surrounded by lush mountains and deep valleys, the temple presented a self-contained landscape scenery (Figures 6–7). Another example is the Guangzhai Temple built in 508 CE. It used to be a residence of Emperor Liangwu. Having the Chishi Rockeries in the front and the Egret Island in the rear, the site stood out as a beautiful landscape. The Jiming Temple, rebuilt and renamed in 1387, makes another example. It was originally built on a backyard during the Western Jin Dynasty period. In 527 CE, the Tongtai Temple was built on the site to facilitate the worship activities of Emperor Wudi of Liang Dynasty, as it was close to the Tai Township. The temple garden, as such, only changed the nature of the structure, without altering the surrounding landscape (Figures 8–9). Second, during the Six Dynasties, most Jinling temples would be built nearing the settlements that had a relatively dense population. Meanwhile, the settlements would find themselves in the areas enjoying accessible waterways and pleasant sceneries (Figure 10). For example, some high ranking officials of the Wu State, including Zhang Zhao and Lu Ji, had their residences built in the vicinity of the Waguan Temple. During the Eastern Jin Dynasty, Wang Dao was an official taking care of the pottery operations in the locality. He granted Monk Hui Li a piece of pottery making land to build a temple in 364 CE. The site sat on a hill with lush greeneries and trees. The hill was named Mount. Phoenix Terrace after the fact that it attracted the perch of flocks of peacock-like lucky birds. During the time of Emperor Wudi of Liang Dynasty, the Waguan Pavilion was built on the hill that attracted tourists' attention for the beautiful landscape. Li Bai, a famous poet at the Tang Dynasty, wrote a popular poem praising the scenes of the Phoenix Terrace, stating that, "The Phoenix Terrace used to be visited by the phoenix. Now one only saw an empty terrace without the phoenix, though the Yangtze River remains flowing eastward. Flowers and weeds buried the desolate trails on the temple premises. How many Jin's royal families had become part of the wildness and ancient mounds? Clouds above the three mountains parted for a sight of blue sky, while the river was divided by the Egret Island into two streams. There is always the case that traitors sit in power as if the floating clouds would sometimes block the sight of sun. I can no longer see Changan (a capital city), feeling lost and depressed." One can vaguely discern from the poem the ties between the ambient environment and the structures. Third, ideologically, the metaphysics, Buddhism and Taoism that flourished during the Six Dynasties shared some commons. They were complementary to one another in exploring the life, and not much contra-ictory to one another in the lifestyle they were pursuing. They were highly consistent in terms of their aesthetic appreciation of landscape. In this context, the development of Jinling temple gardens during the Six Dynasties shall take credit of the accomplishments made by royal and private gardens, especially in the combination of artificial and natural landscape.

As far as the time span is concerned, some three-hundredar's reign of Six Dynasties in the Jinling area bestowed ample time for the development of royal gardens, especially in the Eastern Jin Dynasty. Meanwhile, private gardens enjoyed a steady development during the period of Jin-Southern Dynasties, with temple gardens having a rapid development in the late Southern Dynasties.

4 A poetic view of landscape

Whenever people mentioned the landscape/aesthetics, metaphysics/figures and gardens/life in the Six Dynasties, the most talked about are the Seven Sages of the Bamboo Grove (Figures 11–12) and Orchid Pavilion Gathering. They would always be the original representation of the Wei-Jin Style. The former advocates that"mountains and waters harp their own tunes without the need of music instruments." The Seven Sages of the Bamboo Grove were the representatives of metaphysics at the time. They lived a free and unfettered life, nurturing their virtue and mood amid natural mountains and waters. Unfortunately, they could never be that free and unfettered politically, as they, more often than not, would have a tragic political fate. The latter had its style clearly manifested in Wang Xizhi's *Preface to Orchid Pavilion Poems Collection*: "The site has high mountains and steep peaks, enjoying the presence of lush woods and slim bamboos, surrounded by clear flows of rapids. Here, water serves as a rotating table for wine cups with folks sitting alongside. Folks entertained themselves for one poem per cup of wine while airing their nostalgias, without the company of music band." (Figures 13–14) What a grace and what a clear stream! Both of the two made natural landscape their sites, though with a time interval up to hundred years. One sat on the north of Jinling, and the other on the south. It seemed that no particular attention had been paid to the two in the past. Now one would ask what kind of role Jinling has played in the story of the Seven Sages of the Bamboo Grove and Orchid Pavilion Gathering. Assuming lifestyle and site preference of the two represent a unique literati choice of life, one has to further probe the ties between its influence and Jinling.

From a limited spatial and temporal perspective, Jinling, the political, economic and cultural center of the Six Dynasties, shifted the center of landscape garden development in two phases. Phase one started from the Seven Sages of the Bamboo Grove (240–249 CE) and Orchid Pavilion Gathering (353 CE) that lasted for some hundred years. Jinling not only inherited but also carried forward the aesthetic attitude towards nature, which is of special significance. It was a center of idle talks that eventually attracted metaphysics gathering. It revealed a process that shaped one's perception of nature and associated ideology, as discussed in the preceding two parts of the paper. It made a core of landscape gardening, as it yearned for the idealized personality, ambition, life and feeling. Phase two covered a two-hundred-year period from the grand literati gathering at the Orchid Pavilion in 353 CE to the termination of the Southern Dynasties. The enhanced landscape awareness and the fermentation of landscape garden development in Jinling encouraged the flourishing growth of landscape gardens across the Southern Yangtze River Delta region. In addition to the Jinling garden regime discussed in the third part of the paper, an array of other areas, including Guangling (today's Yangzhou), Wuxing (Huzhou), Jingkou (Zhenjiang) and Wujun (Suzhou), also enjoyed a laudable development of landscape gardens. The Six Dynasties heralded an epoch where landscape gardening started to have a profound change of its connotations with increasingly diversified styles, and where the combination of artificial and natural landscape played out as popular scenes (Figure 15). It was a process that led to the formation of a unique cultural regime and associated basic forms.

From an extended spatial and temporal perspective, the political center had shifted from Luoyang and Chang'an in the north to Jinling in the Southern Yangtze River Delta, and back to Chang'an in the north during the period from the Han Dynasty to the Sui and Tang Dynasties. During the period, the Six Dynasties created a spatial and temporal conjunction that could not be passed around or ignored, as it had forged a cultural link that treated life in a free, unfettered and poetic manner, and brought human souls back to nature. The floating cup practice was first introduced and recorded during 227–239 CE, where Emperor Ming of Wei Dynasty feasted his ministers by setting up floating wine cups along the water channel.[17] In the first lunar month of 329 CE, the royal palace of Taicheng was attacked, and the Eastern Taiji Hall was burned to ashes. The Jianping Garden was built in February serving as the palace. A new palace was put into operation in 330 CE. The emperor and his officials celebrated the event by staging a floating cup feast using the water flowing down from Mount. Zhong.[18] On March 3rd of 353 CE, Wang Xizhi invited his literati friends to have a floating cup picnic at the Orchid Pavilion. The practice became very popular since then. In the first lunar month of 616 CE, Emperor Yangdi of Sui Dynasty ordered to build a palace garden that stretched 12 lies for 16 external palaces and 4 summer pavilions equipped with the streams for floating cups.[19] All these important spatial and temporal developments occurred in Jinling during the Six Dynasties. In addition, the literati gardens, palace gardens, private gardens and landscape gardens that flourished during the Sui and Tang Dynasties are the continuation of the same endeavors stemmed from Jinling during the Six Dynasties (Figure 16). These endeavors and grand practices abruptly came to an end during the Southern Dynasties. However, one still can discern the pursuit, aftertaste, appreciation and passion depicted by the Tang poets in their poems about Jinling's landscape and gardens, from which one may feel the splendid and extended charms, unique aesthetic appreciation and pastoral aura that the Six Dynasties used to have. In one of his poems on the "City of Taicheng", Wei Zhuang deplored, "Misty Spring rains washed up lush greens. Six Dynasties was merely an empty dream like seeing a bird singing in the dream. The most ruthless were the willow trees in Taicheng. They remained standing on the 10-mile embankment amid the misty fogs." In the poem, he spoke not only about the scenes he saw, but also all the more about a realm of intent and a kind of dream journey (Figure 17).

"Living amid the Mountains and Waters of the Six Dynasties"—it speaks about the origin of gardens during the Six Dynasties, and the development of landscape gardening. It demonstrates a unique contribution made by Jinling, and depicts an image of Chinese landscape culture that will last forever.

1

图 1：六朝金陵山水格局依存。中：孙吴都建业图，《金陵古今图考》；照片：陈薇拍摄

Figure 1 : Mountains and waters in Jinling during the Six Dynasties. Picture in middle: Jianye capital map in the Sun-Wu Dynasty, from *Atlas of Ancient and Today's Jinling*; photo by Chen Wei

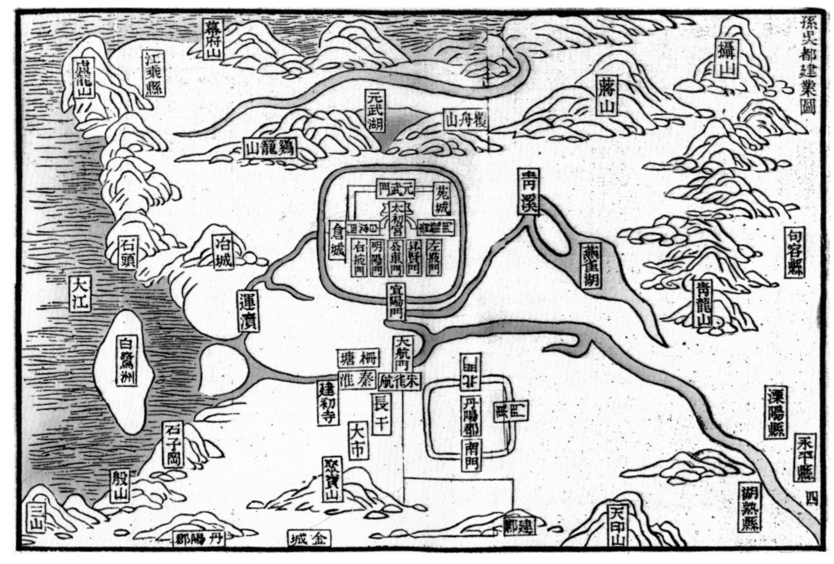

2

图 2：六朝金陵水系图示。底图：孙吴都建业图，《金陵古今图考》；图示：陈薇

Figure 2 : Jinling water systems during the Six Dynasties. Bottom: Jianye capital map in the Sun-Wu Dynasty, from *Atlas of Ancient and Today's Jinling*; drawn by Chen Wei

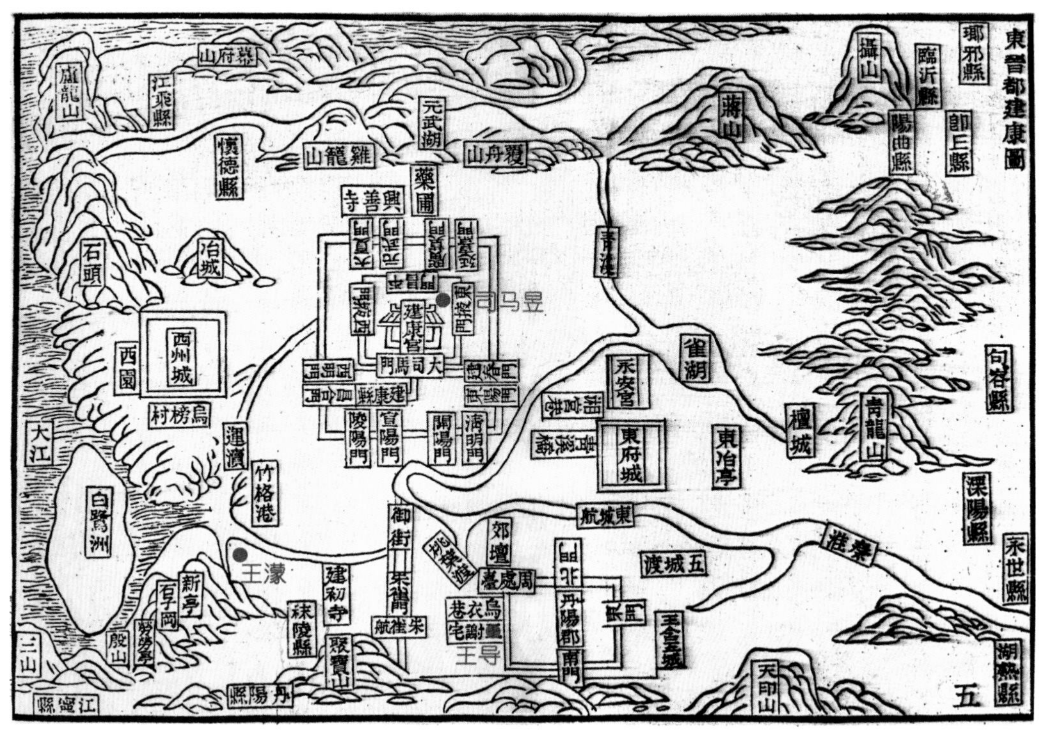

图3：六朝建康玄学清谈中心。陈薇 编制，参见《世说新语》卷上之下“文学”篇描述；底图：东晋都建康城，《金陵古今图考》

Figure 3 : Idle talking centers in the Six Dynasties. Bottom: Jiankang, the capital city of Eastern Jin Dynasty, from *Atlas of Ancient and Today's Jinling*; drawn by Chen Wei

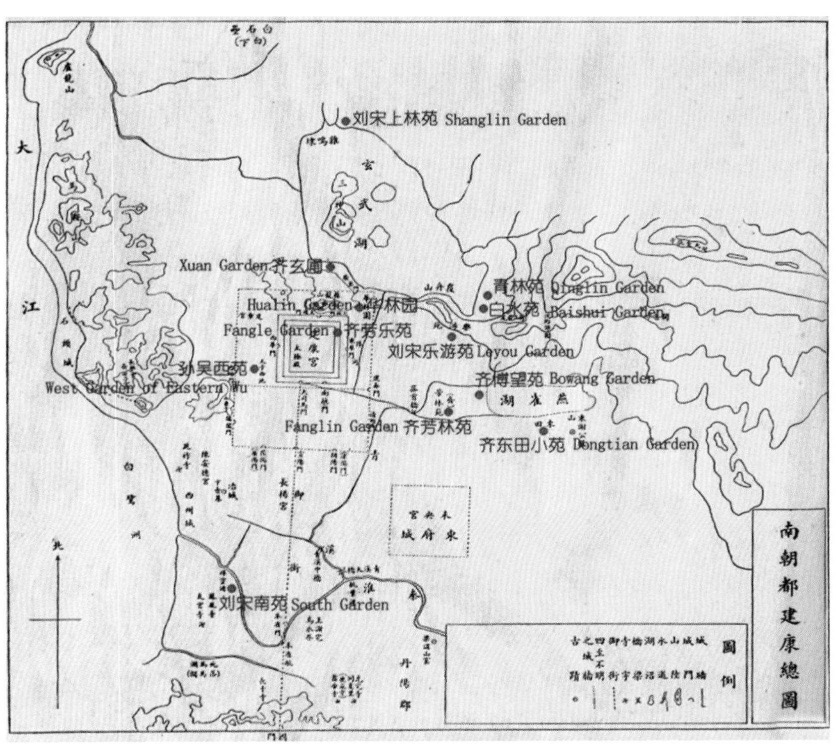

4

图 4：六朝金陵主要皇家园林和私家园林分布示意图。陈薇编制，底图：南朝都建康总图，朱偰，《金陵古迹图考》（上海：商务印书馆，1936 年）

Figure 4 : Distribution of main royal gardens and private gardens in Jinling. Bottom: Jiankang, the capital city in the Southern Dynasties, from Zhu Xie, *Atlas of Jinling Ancient Relics* (Shanghai: Shangwu yinshuguan, 1936); drawn by Chen Wei

5

图 5：六朝木屐，南京集庆路颜料坊出土。南京六朝博物馆藏
Figure 5 : Xie's Shoes, a cultural relics of the Six Dynasties. Collected by Nanjing Six Dynasties Museum

6

图 6 : 六朝胜迹栖霞寺景。陈薇拍摄
Figure 6 : A scene of Qixia Temple. Photo by Chen Wei

7

图 7 : 南朝开凿的栖霞石窟群。陈薇拍摄
Figure 7 : A scene of Qixia Grottoes. Photo by Chen Wei

8

图 8：同泰寺遗址上的烟雨楼台鸡鸣寺（明以后名鸡鸣寺）。王建国拍摄
Figure 8 : Tongtai Temple (Jiming Temple in the Ming Dynasty and after). Photo by Wang Jianguo

9

图 9 : 同泰寺遗址上的鸡鸣寺近依台城。王建国拍摄
Figure 9 : Jiming Temple and Nanjing City Wall today. Photo by Wang Jianguo

10

图 10：小九华山（覆舟山）周围风光旖旎，六朝时建有乐游苑和寺庙。陈薇拍摄
Figure 10 : The surrounding scenery of Xiaojiuhua Mountain (Fuzhou Mountain) is beautiful. During the Six Dynasties, there were amusement gardens and temples. Photo by Chen Wei

11

图 11：上：模印砖画“竹林七贤”，南京六朝博物馆藏；下：“竹林七贤”拓本
Figure 11 : Up: “The Seven Sages of the Bamboo Grove” on brick carving, collected by Nanjing Six Dynasties Museum; bottom: the rubbings of “The Seven Sages of the Bamboo Grove”

12

13

14

图 12：《兰亭图》。（明）文徵明绘。辽宁博物馆藏
Figure 12 : *Orchid Pavilion*. Drawn by Wen Zhengming (Ming). Collected by Liaoning Museum

图 13：兰亭历史环境：崇山峻岭，茂林修竹，清流急湍，映带左右。陈薇拍摄
Figure 13 : Orchid Pavilion and its historical environment: high mountains and steep peaks, the presence of lush woods and slim bamboos, surrounded by the clear flow of rapids. Photo by Chen Wei

图 14：兰亭曲水流觞遗迹。陈薇拍摄
Figure 14 : Relic traces of floating cup practice. Photo by Chen Wei

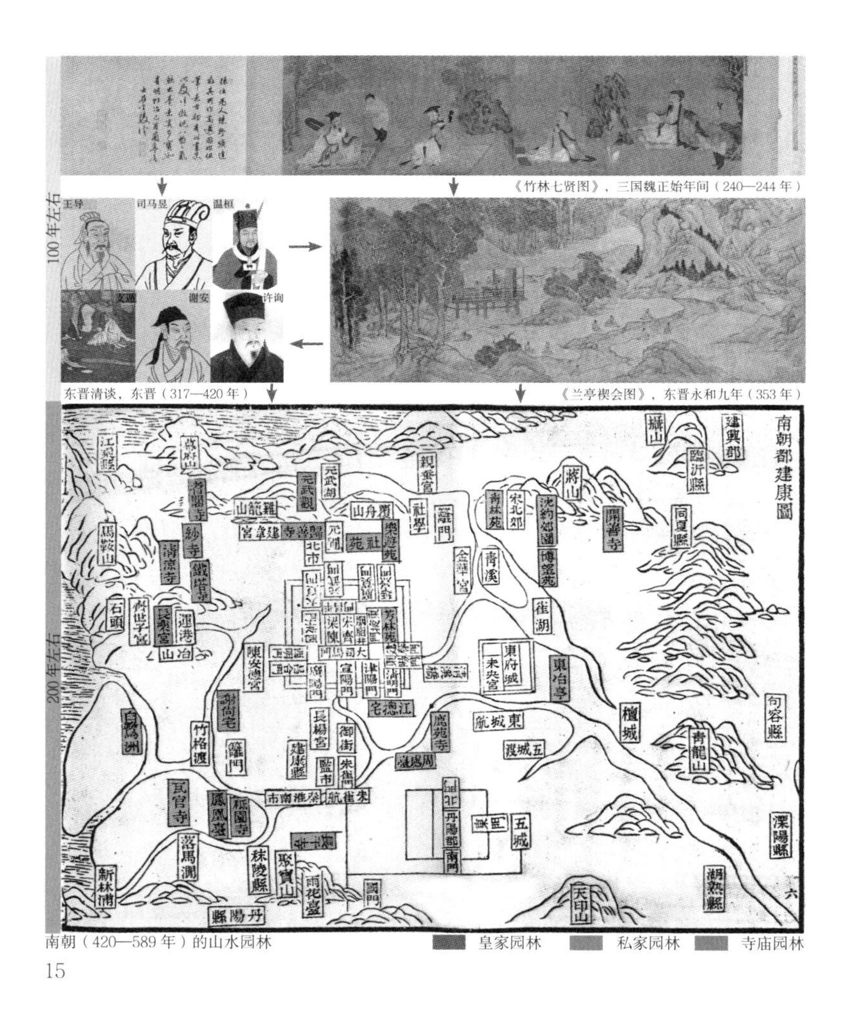

15

图 15：作为中心的金陵地与江南山水园林形成的时空关系图。陈薇编制。底图：谭其骧主编，《中国历史地图集》第 4 册 东晋（北京：中国地图出版社，1982），第 5-6 页；引用图片：（唐）孙位《高逸图》，为《竹林七贤图》残卷，上海博物馆藏；（明）文徵明《兰亭修禊图》，北京故宫博物院藏

Figure 15 : The spatial-temporal relationship between Jinling Land as the center and Jiangnan landscape gardens. Drawn by Chen Wei. Bottom: Tan Qixiang, ed., *Zhongguo lishi ditu ji* (Beijing: Zhongguo ditu chubanshe, 1982), 4.5–6; Quoted picture: Sun Wei (Tang), *Gao Yi Tu*, the remnant of the painting of *Seven Sages in the Bamboo Grove*, collected by Shanghai Museum; Wen Zhengming (Ming), *Orchid Pavilion Repair Painting*, collected by Beijing Palace Museum

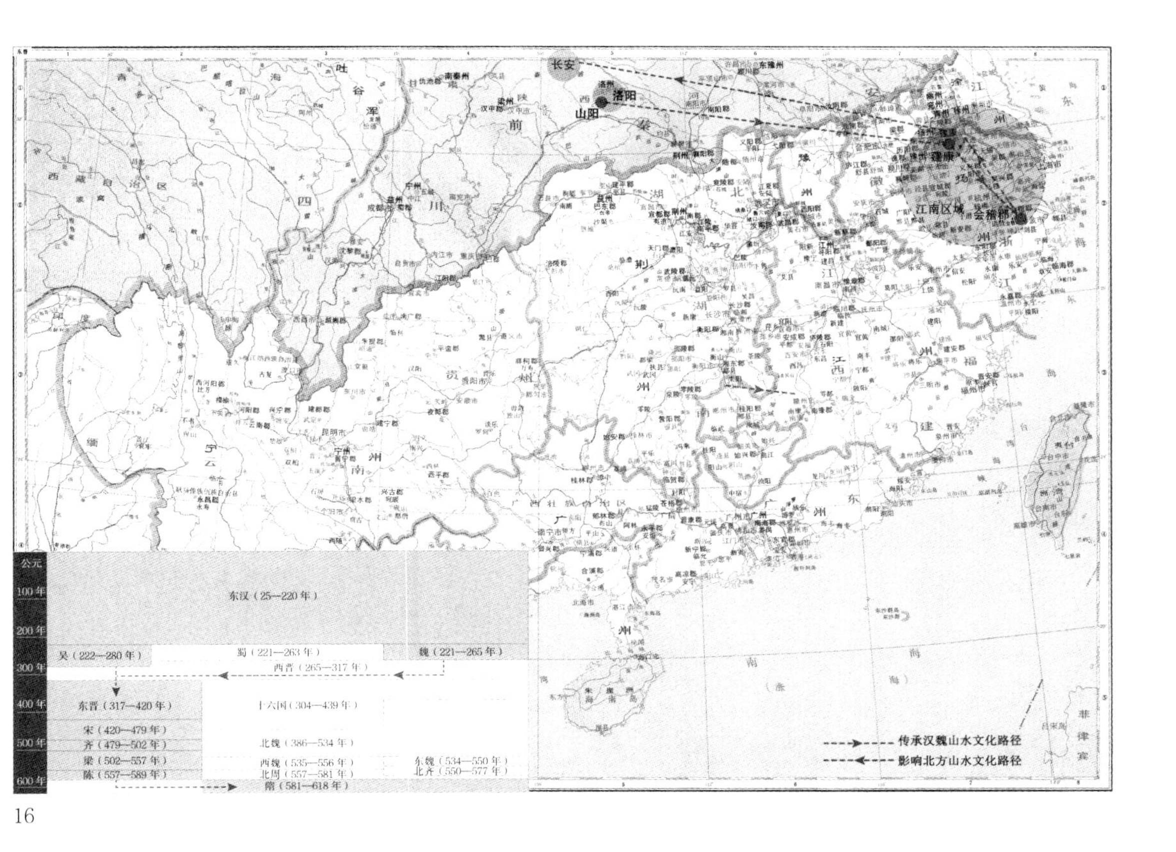

16

图 16：金陵跨越时代对山水园林影响的时空关系图。陈薇编制；底图：谭其骧主编，《中国历史地图集》第 4 册 南朝（北京：中国地图出版社，1982 年），第 42–43 页

Figure 16 : Temporal and spatial relationship of Jinling's influence on landscape across the ages. Drawn by Chen Wei; bottom: Tan Qixiang, ed., *Zhongguo lishi ditu ji* (Beijing: Zhongguo ditu chubanshe, 1982), 4.42–43

17

图 17：金陵十里堤风景依旧。陈薇拍摄
Figure 17 : Unchanged Jinling 10-li embankment scene. Photo by Chen Wei

Landscapes for Travelling, Sightseeing, Wandering or Dwelling

Minna TÖRMÄ

This essay takes as its starting point observations by two individuals who seem to be worlds apart not only geographically but also temporally: Osvald Sirén (1879–1966) and Guo Xi (11th century). At the beginning of his book *Gardens of China* (1949), Sirén proposes a close connection with the viewing of a long landscape handscroll and the enjoyment of a garden (Figure 1). I will explore the ways in which this analogy might or might not be tenable. I will also consider this in the light of Guo Xi's well-known division of landscape paintings into four categories: those which are suitable for travelling, sightseeing, wandering or dwelling.

However, before discussing these ideas, brief introductions as to who Osvald Sirén and Guo Xi were are in place. Osvald Sirén was Finnish-Swedish art historian and one of the early 20th century pioneers of Chinese art studies in the Western academia. My book *Enchanted by Lohans*, discusses the middle years of his career, from 1910s to late 1930s, when he turned from a specialist of Italian art into one of Chinese art.[1] During those years he covered Chinese architecture, sculpture and painting in his publications. The garden books, first one on Chinese gardens and then another one on Chinese elements in European gardens, were published in the late 1940s, first in Swedish.[2] They, particularly the first one, were written under the dark years of the Second World War, when Sirén was isolated in Sweden and had very few contacts beyond it, particularly to colleagues in Europe.

During these depressing times, Sirén turned to memories of his wanderings around Chinese and Japanese gardens and reworked those memories into the garden books, complementing the recollections with research, translating Ji Cheng's *Yuanye* and so forth. During his sojourns in East Asia he had photoed the sites carefully and so *Gardens of China* is copiously illustrated with his photos and the book remains valuable as documentation. The photos were taken first in 1918, then 1921–1923, 1929–1930 and 1935. He was an avid photoer so he carried his camera with him everywhere and we can thank him for documenting not only gardens, but also Chinese architecture, particularly in Beijing, and sculpture at the cave sites and beyond.

Guo Xi may not need much in the way of introduction, as he is one of the great masters of Chinese landscape painting in the Northern Song Dynasty (960–1127). He was Emperor Shenzong's (r. 1067–1085) favourite painter and so his paintings adorned the halls of the imperial palaces. His most famous surviving painting is *Early Spring* (1072) in the National Palace Museum (Taipei) collection, but a handscroll from the collection of the Metropolitan Museum of Art (New York) is also an important example of his work and more pertinent to the discussion in this essay: *Old Trees, Level Distance* (Figure 2). It is a short handscroll, which has recently been the focus of an in depth study by Ping Foong: *The Efficacious Landscape: On the Authorities of Painting at the Northern Song Court.*[3]

Presenting a level distance view of a river landscape, the scroll opens with a hazy view of the river with two fishermen in their boats, and beyond them in the distance two travellers with a donkey disappear in the misty landscape. As one unscrolls the painting further, a pavilion comes in view: it is surrounded by a cluster of trees and rocks. A pavilion could be seen as a marker of a garden (turning a landscape into a garden) or of scenic point of view within the landscape. Here the latter interpretation is corroborated by the group of people approaching the pavilion. One of them, a servant, is carrying a long narrow parcel which contains a *qin* and others are bringing provisions for a picnic. From the pavilion they can explore the beautiful scenery while they enjoy refreshments and music.

The text *Lofty Message of Forests and Streams* introduces us to

Guo Xi's thoughts on painting.[4] Guo Xi emphasised landscape painting's potential to take us on a journey or *woyou* "travelling while lying down." He was not the first one to expound this idea, for example Zong Bing (375–443) discussed this already in the fourth century, but Guo Xi elaborated on it further by distinguishing four modes of speed for *woyou*.[5] Depending on the type, a landscape painting would be suitable for travelling, sightseeing, wandering or dwelling.[6] Handscroll as a format is most befitting to mind travel particularly when the painting is constructed with a sense of clarity and precise knowledge of travel methods. Guo Xi's *Old Trees, Level Distance* provides a platform to what apparently looks like a short journey as the painting is only 104.4 cm long and can be easily viewed as a single scene. Nonetheless, the way he has arranged the composition and furnished the details leads the viewer to spend more time in it than simply rushing through: the viewer is prompted to join the procession to the pavilion and spend a leisurely day in their company there.

Viewing a longer handscroll will bring Guo Xi's ideas to clearer focus. *Streams and Mountains without End* by an unknown Northern Song painter takes the viewer along on a long journey (Figure 3).[7] The painting opens up with scenery of a river valley and the mood is misty and expectant. A path leads the Traveller (internal viewer) behind mountains.[8] Once the path re-emerges, the Traveller has arrived at a gate of a courtyard house (a country house with thatched roofs and bamboo fence) and beyond the buildings is a mountain terrace. Following a route past the gate, the Traveller may notice a pavilion on a hillock. The external viewer will see also a house on a cliff by the river in front of the pavilion; however, this may be unknown to the Traveller. The path broadens and the Traveller arrives at a promontory: another pavilion juts out on it and there is also a docking place for boats. One boat is just departing and somebody has come to see off the person boarding the boat. Other figures have gathered on the promontory as well and one of them is carrying a long narrow parcel, which might be a *qin*, like the one we saw in *Old Trees, Level Distance*. The Traveller may decide to hop on a boat and continue the journey along the river enjoying the scenery at a distance, but he may equally choose to resume the journey on foot by back tracking his steps and turning right on the path which follows along the side of the mountains.

This path leads inwards in the picture and disappears into the mountains. The Traveller is at crossroads here and has to decide whether he wishes to vanish along it into his imaginary wanderings or to turn left to see what the painter has in store for him. This road is wide and turns into a bridge steering over a rivulet from the mountains joining the wider waterway. Behind the bridge on the shores of the rivulet is a group of buildings and from the bridge he sees across the bay the pavilion he visited earlier. After the bridge the path recedes from view behind the huge boulders on the shore and reappears along a wider stream feeding into the river. The path is built on poles along the mountainside and the Traveller sees no option to cross the stream but instead wanders into the mountains. The spot is not shallow enough to wade through otherwise the artist would have painted people doing so as an example to emulate. While the Traveller makes his way through the mountains, the external viewer enjoys the landscape scenery from a distance.

In case the Traveller had decided to skip this arduous journey on land and had stepped on a boat he would have reached the next cluster of habitation faster. Arriving from water, a boat would reach a promontory with a pavilion and could dock there. Once on the shore again, the Traveller's journey would continue behind the boulders and the clump of trees to an opening surrounded by thatched houses. However, a top of a gateway is just visible further back and the Traveller who went to explore the land route would be entering the village through that gate. The village provides a place of rest as well as a converging point for the two modes of travel. It is sheltered and tucked in the embrace of the range. The route leads now out of the village over a stream: on the bridge a figure on horseback and another on foot are coming towards the Traveller. But again an attentive eye sees another option in the form of a path, which has led from the village square behind more boulders. It guides the eye deeper in the mountains veiled in mist.

The road along the main waterway, however, offers an easier way forward and a possibility to join a lone traveller—maybe stop with him at the pub on the roadside, the modest house with a flag outside. The main route will take the Traveller to a temple, its towers rising high behind the pinnacles. But passing that option and a searching for an alternative path steers the eye again into the depths of the mountains. For the external observer, a large mountain mass dominates the view at this point. There may be a shallow spot to the stream, which is trickling down to the main river, because when the road comes into view again by the shoreline more travellers are pushing ahead. But the effort to see where the road is leading to is halted. Instead, the eye discovers two fishing boats tucked into the safety of the meandering shoreline. A fisherman stands on one of them but the other one seems quiet. The Traveller could ask the fisherman to ferry him across the water but as the eye roams over past the mountain range and over the valley landscape no paths or roads for the continuation of the journey are visible. It is time to track back the steps or settle down fishing.

How does this scroll fit into Guo Xi's categories of travelling, sightseeing, wandering or dwelling? All types are about being in the landscape, experiencing from within; each refers to a type of temporal rhythm changing from a relatively rapid movement to staying in one place. As a viewer opens a handscroll, he or she begins by noting the general scenery of the opening frame (for example a river valley) before zooming into details. The viewer is looking for an entrance point for the exploration of the landscape from inside while observing the representation of scenery from outside at the same time. The intriguing aspect of *Streams and Mountains without End* is that the painter has offered more than one option for the viewing tempo through the scroll. There is what we could call a main route along the river following the lower border of the painting and where a "shortcut" can be made on a boat. This would accord with the aspect of travel in Guo Xi's terms. However, a slower unfolding of the landscape is suggested as well with the diverging paths inwards into the image. Following these diversions, which are parts of the route of the internal viewer, the external viewer may loosen himself in his memories of his own wanderings among mountains and valleys. Throughout the viewing process there is interplay between the Traveller and the external viewer. In a sense it might be argued that the external viewer has a more privileged position and he is in command of the larger view in each frame, but it is the internal viewer who provides the lifeline for the mind to wander.

In my book *Landscape Experience as Visual Narrative: Northern Song Landscape Handscrolls in the Li Cheng-*

Yan Wengui Tradition (2002), I was focusing on a handscroll to Yan Wengui (ca. 967–1044).[9] The painting, *Pavilions and Mansions by Rivers and Mountains (Jiangshan louguan tu)*, is in the collection of the Osaka Municipal Museum of Art in Japan. One of my main arguments in the book is that even if a handscroll does not seem to have a story to tell, when we view it, a story begins to build up in our minds during the viewing process. Clusters of detail tend to slow down the viewing process and this is how the painter can create the narrative rhythm, which is the basis of Guo Xi's categorisations. His discussion on the essentials of landscape representation emphasises clarity and rationality. In order for a landscape painting to be successful it should fulfil its main objective, that is, to assist the viewer to travel in his mind. The main reasons for the desire for mind travel were old age, illness, relaxation in general and meditation. This trust in the power of the mind leads us to the Chinese concept of *li*. The Northern Song philosopher Cheng Yi (1033–1107) explored how *li* was linked to dreams, memory and so forth:

> What the mind "penetrates when stimulated" is only the *li*. The events of the world as it knows them either are or are not, irrespective of past and present, before and after. For example, whatever is perceived in dreams is without form; there is only its *li*. If you say that dreams are concerned with such things as forms and voices, these are ether.
>
> Within heaven and earth, what is simply is. For example, what a man has experienced, has seen and heard ... one day after many years he may recall it, complete in his breast. Where has this particular *Tao–li* been located?[10]

Li determines how matter and objects are structured and it could be called the internal organising pattern in everything that exists. "What the mind 'penetrates when stimulated' is only the *li*." This approach makes it clear that landscape paintings must follow and manifest the same *li* as the actual landscape. If it deviates from this principle a landscape painting is doomed to failure in assisting with *woyou*.

Since it is advisable to study and describe handscrolls by unrolling them frame by frame rather than spreading them open in their entirety, then it would also be more appropriate to analyse a garden in the form of a walk-through instead of looking at an overall plan.

As mentioned above, at the beginning of his book *Gardens of China*, Sirén proposed a close connection with the viewing of a long landscape handscroll and the enjoyment of a garden.[11] He emphasised that a garden cannot be viewed from a fixed point; instead, a Chinese garden consisted of more or less separated fragments, which followed one another in an organic fashion forming in the end a unified composition. At this point of his introduction he is also recommending a walk-through at a leisurely pace as the basis for an analysis of a garden's structure. However, later on in the book, when Sirén explores individual gardens he does not engage in detailed descriptions of his routes through the gardens. Instead, his discussion of Wangshiyuan for example is only a paragraph long.[12] Wangshiyuan in Suzhou is a small garden. It invites the visitor to linger, to walk slowly and to pause once in a while to admire view and details. It is densely packed with buildings and paths and corridors meander and connect different parts. It is a veritable labyrinth and to me it is a perfect example of a garden, which unfolds slowly. It is organised around a pond: as one walks around, the pond sometimes comes closer, sometimes recedes and occasionally disappears from view.

However, the journey through this garden begins through three halls following each other in a formal manner along a central axis. It is only when the visitor has passed through these halls that the actual garden labyrinth begins. If at the end of the three halls, one turns through a door on the left, one arrives at a pavilion on the shore of the pond. As this is the actual start of the garden journey, it is a suitable spot for pausing and absorbing the view spreading out in front of you. This is where you step into the landscape. What immediately draws attention is a cluster of water lilies and "The Pavilion of Moon Arriving and Breeze Coming" straight across on the other side. Only a corner of it is visible in the photo (Figure 4).[13] A covered corridor leads from the pavil– ion to the left to a building called "Washing Cap-Strings" which features a balcony built over the water and an entrance to a rock cave.[14] Glancing around toward the right of the pavilion the Traveller notices a zig-zag bridge which gives access to it. From the point where he stands this seems to be the most obvious route to follow. But in order to reach the bridge, he must turn right and walk along the balconies of the buildings framing the pond at this point and through a moon door. Once on the bridge the Traveller looks to the right at "The Hall for Viewing the Pine and Seeing a Painting" and to the left across the pond how the angle of the view towards "Washing Cap-Strings" has changed as he has moved along before he steps up to the pavilion. It is a perfect spot to rest the feet and admire the scenery around the pond. The entrance to the garden is now on the opposite side and the trees, plants and rocks create a pleasing picture against the white washed wall of the building behind them. If this were the night of the mid-autumn moon, the full moon would be directly visible above, thus the name of the pavilion. After a while the Traveller continues the tour along the covered walkway leading to "Washing Cap-Strings," which had caught the attention in the beginning. From there one can again look back to where one entered the garden (Figures 5–6) and see the door and the building through which one reached the zig-zag bridge. A glance towards the left gives a side view of the pavilion (Figure 7).

An entrance to a cave constructed of rocks to his right invites the Traveller to enter and the opening leads to a slightly larger building than the "Washing Cap-Strings." Behind it he discovers more small pavilions and the smallest one is called "The Lute Room." This area is a kind of a dead end though a pleasing corner to linger for a while, but the route leads back towards the main pavilion by the pond. Passing it again the Traveller notices an opening on the left and enters a large open space. The space is bordered by white washed walls, which create a paper-like white background for composition of rocks and plants, the result being like a detail from an ink landscape close-up with grey washes of ink and a bit of green colour as accent (Figure 8)—similarly to what was observable when looking from the pavilion over the pond towards the entrance. A pavilion-like structure has been placed against the long wall opposite to the doorway. It is inhabited by a large rock so the pavilion functions both as a display platform and as a place of rest since benches border the rock on both sides (Figure 9). Like the rocks and plants, the red of the pavilion with the grey rocks creates a pleasing picture against the white wall. The rock, which is the focus of the display, is a Taihu stone: one of those famous stones, which were avidly collected by emperors and formed the most valued treasure of many a garden. Emperor Huizong (r. 1100–1125) was one of the most enthusiastic collectors of these rocks and immortalised one in his painting *Auspicious Dragon Rock.*[15] "Late Spring Studio" at the one end of the courtyard is worth exploring and from the side door there the Traveller can step out and move to the next hall which is called "The Hall for Viewing the Pine and Seeing a Painting"—he had gotten a glimpse of it already earlier when on the bridge. Its doors give out to the direction of the pond; in fact, the whole wall can be opened up to let the scenery and light in. So far the journey has more or less circled around the pond: the paths have moved along its shores or made short diversions into spaces invisible to it, yet always returning back. The Traveller returns to buildings, which he passed through at the start of the journey when he was focused on reaching "The Pavilion of Moon Arriving and Breeze Coming." He enters again through the moon door and realises that there is a whole area yet to be explored in the opposite direction of his earlier route. The journey seems to find its final stop at "The Hall of Ascending to the Clouds," as if hinting that this is the culmination of the earthly journey. The garden design is carefully constructed of pavilions, halls and courtyard space cells following one another in a labyrinth like manner and in the end the Traveller loses sense of direction and time. Garden is viewed from the point of view of the Traveller (internal viewer) only. What would the result be, if one were to take snapshots often along the way and put them together for a panoramic view?

While one wanders around in a Chinese garden a number of details catches the attention: there are windows to inaccessible spaces (Figure 10) and openings, which hint about further spaces for exploration (Figure 11). While sitting down one may be gazing at a miniature landscape (Figure 12) and sometimes it is worthwhile to stare at what is happening under your feet (Figure 13). The scenes framed by a moon door (Figure 14) or by a window are like a picture within a picture, a method of representation, which is familiar in Chinese painting.[16] In painting, this approach uses screen representations as a pictorial device to add layers of meaning to paintings. Most of the time the painted screens are just what they are: pieces of furniture with a fashionable decor. However, there are also a number of paintings in which the image on a screen is like an opening to a new world or to an added dimension. Examples of the latter kind are, for example, Liu Guandao's (active 1279–1300), *Whiling Away the Summer* and a Ming version of *The Double Screen: Emperor Li Jing Watching his Brothers Play Weiqi.*[17] Through the screen images in these paintings we can enter another moment or world in our minds. The framed window landscape (Figure 10) opens up a view to a landscape beyond our space, which is at the same time physically unattainable. The carved lattice work of the window enhances the picture-world quality here. A moon door as a frame is invested with more possibilities for interpretation as it can be likened to a cave entrance and thus is a boundary between two kinds of worlds: the real and the imaginary.

In searching descriptions of Wangshiyuan in Western literature on Chinese gardens, the writers seem to be rarely taking this wandering approach to its description, possibly because it does get rather complicated. My description above missed many details and potential diversions. Besides, the Traveller constantly gazes backwards, particularly around the pond, which makes an attempt to describe the garden experience more challenging. A more common way is to talk about gardens in general terms and arrange chapters under particular themes such as water, stones, poetry, architecture, etc. In these cases, references to Wangshiyuan are scattered over several chapters, like in Sirén's *Gardens of China.* In *A Chinese Garden Court: The Astor Court at the Metropolitan Museum of Art*, Wangshiyuan receives a more comprehensive treatment as a whole: the reader is offered a plan of the garden and descriptions circle around the different corners of the garden in relation to the plan.[18]

Chen Congzhou (1918–2000) proposed in 1984 that gardens should be divided into two main types:

1. "In-position garden viewing" and his main example for this is Wangshiyuan; he further explains that this refers to "lingering observation from fixed angles," meaning that "there are more visual points of interest to appreciate from fixed angles."

2. "In-motion garden viewing" and the main example for this is Zhuozhengyuan; and this refers to "moving observation from changing angles," meaning that this type of garden "demands a longer 'touring' vista."[19]

If we think about the two garden types proposed by Chen Congzhou, we could develop the analogy between a handscroll and a garden. Could we argue that landscape paintings, which are for travel and sightseeing, would accord with the category of "in-motion garden/landscape viewing," if we modify Chen's expression? And then those, which are for wandering and living, would be about "in-position garden/landscape viewing"? It could also be pointed out that gardens like Wangshiyuan are for wandering and living, whereas Zhuozhengyuan is more suitable for travelling and sightseeing.

The analogies to pictorial representations are not limited to the handscroll format. Alfreda Murck and Wen Fong propose in their discussion of the Astor Court that "[f]rom any given vantage point in the Chinese garden, say from the veranda of a hall, the visitor sees a fixed view as in an album leaf, but as he moves from one point to another, as along a walkway, the scenery unfold as it does in a *handscroll*."[20] However, the view from a given vantage point could also be compared to a handscroll as viewed by an external viewer. This brings me to the question I posed earlier: What would the result be, if one were to take snapshots often along the way and put them together? For example, we could place the photos in Figures 3, 4 and 5 next to each other and that would create a panoramic view when sitting by the railing in the "Washing Cap–Strings." I would argue that the views are rarely as fixed as the above quote or Chen Congzhou suggest. The eyes will continue roving about and the head turning around even when the rest of the body is still.[21] Earlier, I also proposed that there might be no external viewer in the garden experience, but it would perhaps be more accurate to think of the Traveller moving between the positions of an internal and an external viewer while he journeys through a garden. The Traveller would be in the position of an internal viewer while wandering and moving about and in the position of an external viewer while stopping or sitting down to absorb the prospects.

A handscroll by Cheng Zhengkui (1604–1676), *Dream Journey among Rivers and Mountains* (no 150, dated 1662) provides a fascinating counterpoint to the discussion on the relationship or a garden and a handscroll (Figure 15). Opening the scroll, the first view shows a group of buildings tucked comfortably into the landscape behind a hillock. On the hillock grows four twisted trees. Beyond it a river landscape winds into distance. Left of the hillock, a bridge crosses the river, so we know that there is a path there for the Traveller to follow. On the opposite bank more buildings—only their roofs are barely visible—are snuggly protected in a pocket created by boulders and the rising mountains. From that point, a view opens towards a waterfall cascading down a mountainside on the opposite side of the stream and a path winds upwards curving to the left and disappearing from view. This is a moment of slight disorien–tation, but scrolling further, after the mountain range a web of paths in valley comes into view. While deciding where to go, the Traveller is suddenly confronted by a huge Taihu rock in the middle of the landscape almost blocking his route. It certainly catches attention and both the Traveller and the external viewer pause facing it while contemplating this unexpected feature. There are other anomalies regarding scale at this point too. On the foreground, there are what appeared to be at a first glance mountains with spruces growing on top of them. However, these look like miniature versions of great mountains within the overall scale of the image: more like rocks planted with spruce seedlings as if on a tray landscape. The anomalies may be more striking to the external viewer, though the Taihu rock may seem inexplicable to both. From the point of view of the Traveller, the miniature-like landscapes could be just mountains in the distance. But before elaborating further on the reversals of scale and elements of surprise, it will be worthwhile to finish the journey through this painting.

So far, there has been no visible human presence and only the houses have indicated that the landscape is habitable. As the Traveller passes the Taihu rock and turns left along the mountainside (the traveller could also have taken the path which disappears behind the mountain), he catches up with two scholars wandering along the path. The path descends and a bridge leads over a river. The Traveller notices houses upstream along the riverbanks: however, no paths seem allow access to that area. The route, which is laid out for the Traveller, continues first along the foreground and then turns upwards and into the mountains. The external viewer loses the sight of it and is, instead, lead to explore more building squeezed in between the rising mountains. A glimpse of a pavilion roof peeks out behind a cliff top. Just beyond it, the path curves back into view, where a boat is on the river and a person is coming towards the Traveller. Passing him, the Traveller circles around a steeply rising unapproachable mountain.

Once on the other side of the mountain, the Traveller stands on a promontory and sees a house on top of a cliff jutting over water. An old gnarled and bent pine grows at the foot of the cliff. A broad expanse of water dominates the landscape at this point and the distant mountains are shrouded in mist. Two scholars are standing on this side of the shore on the riverbank. One of them is pointing upwards, drawing attention to the flight of birds—possibly geese. To the left from them, on the riverbank another Taihu rock has been given a prominent position, though not quite as grand as the previous one. After this point the landscape remains fairly remote and the Traveller would need a boat to get access to the nearby islets or the mountains in the distance. There is only a very narrow strip of the shore visible in the foreground, but that is the only route forward for the time being. Two scholars are engaged in a discussion further along the shore. At this point the mountains on the opposite side are moving closer: a stream is trickling down and there is a bridge over and a path along it. Having gone past the two scholars and glanced at the stream, the Traveller continues behind a boulder and reaches a long bridge where he may join a group of three men. The current looks swift underneath. Once over the bridge, the mountains become more impenetrable. The external viewer encounters mountain forms, which stretch from the bottom to the top of the image. Scrolling further, the external viewer senses that this point was a rupture and that after this point the landscape becomes inaccessible to the Traveller whose route within the landscape is blocked at the wall of this massive mountain. In addition, for the remaining scenery, there is a similar reversal of scale as around the first Taihu rock. The landscape elements along the lower edge of the painting are small, as if they were distant and vice versa, until to the very end of the scroll when a move towards normal scaling among mountains and waters returns.

Cheng Zhengkui titled his painting as a "Dream Journey" and he is known to have painted numerous handscrolls with the same title. He meant them as gifts to his fellow colleagues at the newly established Manchu court in Beijing and the one discussed in this essay is numbered 150.[22] With this knowledge in mind, the changes of scale, awkward forms and a sudden appearance of a gigantic Taihu rock can be interpreted as common features of the dream world. However, I would like to propose an alternative interpretation particularly for the handscroll number 150. The painting is a landscape and yet there are details, which give a sense of wandering in a garden. The Taihu rocks are the most obvious ones, but the changes of scale have a similar effect. In addition, the persons in the painting, mostly in groups of two or three, seem to be strolling about in a leisurely manner, engrossed in a conversation or admiring the scenery. Gardens are landscapes and when wandering in Wangshiyuan, for example, the Traveller is expected to interpret the changes of scale from tray landscapes to artificial mountains created from rocks as features of a real landscape. The interpretation of Cheng Zhengkui's painting as an imaginary garden actually relates well to what Wai-yee Li discusses in her article "Gardens and Illusions from Late Ming to Early Qing."[23] Her research shows that "[t]he trend of writing about illusory gardens arguably started in the late Ming" and that at the same time there was a fascination of dreams and illusions in those writings.[24] This coincides with Cheng Zhengkui's life as he was born in the late Ming, saw the collapse of the Dynasty and then chose to serve the new Qing Dynasty. A number of the gardens which Wai-yee Li explores existed only as texts and had, for example, such names as "Non-existent Garden" and "Imagined Garden."[25] These texts provide dream-like imaginary journeys through gardens where everything is not necessarily as it seems. There is a tendency to cling to the imaginary/fictive rather than the real/concrete and that the uncertain and chaotic historical circumstance spanning most of the seventeenth century led scholars—and others too—to relinquish materialism and seek refuge in the imaginary.

The relationship of a garden design and a handscroll is more complex than Sirén or others have allowed as I hope to have shown in this essay. The comparison is not without its merits; however, we should be more sensitive to the breadth of the variety of nuances in both art forms. We could also consider the position of the external viewer as that of the painter or the garden designer. The viewer of a handscroll or the visitor to a garden oscillates between two points of view, that of the creator and that of the Traveller. Of course there is more to the landscapes and gardens than the Traveller's narrative. The transformations created/described by brush and ink form another aspect of narrative in the paintings and also in the gardens. There is the interplay between surface patterns and what they represent (trees, boulders, cliffs etc.). However, that discussion is beyond the scope of this essay.

In conclusion, we might consider the garden aspect of an earlier landscape painting, *Tea Drinking under the Wutong Tree*, a short handscroll attributed to Tang Yin (1470–1523).[26] It can be easily viewed as one unified image; nonetheless, because it is in a handscroll format, there is a certain kind of anticipation at the beginning. When you start unscrolling you do not know how long the painting will be. Here the first thing you see (apart from the red seals in the lower right hand corner) is a bridge over a stream inviting you to step on it. At the same time, you savour the manner in which the bridge and the rocks nearby are painted. As you gaze upstream, you notice a young boy washing a bowl there. Unscrolling further, you encounter a scholar sitting on a platform holding a teacup and opposite a monk seated on a bamboo chair sipping tea from his cup. A servant is standing under a wutong tree having just prepared the tea on the bamboo stove. Though it seems that they are on an excursion, they might actually be in a garden imagining that they are out among mountains and streams. You might scroll even further past them seeing that the painting ends soon beyond the banana plants. This invites the viewer to join in and have a cup of tea.

可行、可望、可游、可居之山水

米娜·托玛

（梁洁 译）

本文跨越时空考察了喜龙仁（1879—1966 年）和郭熙（十一世纪）这样两位人物，并以之作为论述的开端。在其著作《中国的园林》（1949 年）的开头，喜龙仁提出了观看山水长卷与游园的密切关联（图 1）。我将探索这种类比可能成立或不成立的方式。我也会在郭熙对山水的四种分类中思考这个问题，即可行、可望、可游、可居之山水。

在讨论这些想法之前，有必要简要地介绍喜龙仁和郭熙是谁。喜龙仁是芬兰－瑞典籍艺术史学家，是西方学术界在二十世纪早期研究中国艺术的先驱之一。我的书《痴迷罗汉》讨论了他学术生涯的中期，即二十世纪十至三十年代晚期，彼时他的专业领域正由意大利艺术转向中国艺术，其论著覆盖中国建筑、雕塑、绘画。[1] 他的有关园林的书籍于四十年代晚期以瑞典语发行，其中一本是关于中国园林的，之后一本则是关于欧洲园林中的中国元素的。[2] 他的这些著作，尤其是第一本，均写于第二次世界大战期间的黑暗年代，彼时的喜龙仁在瑞典与世隔绝，基本无法联系到他在欧洲的同事。

在这段压抑的时期，喜龙仁开始回忆他在中国园林和日本园林中的漫游体验，并将这些回忆重新组织、撰写园林著作，对于记忆的不足以研究作补充，翻译计成的《园冶》等等。在东亚短暂的逗留期间，他仔细地拍摄园林，故而《中国的园林》中的插图有丰富的照片——这本书也因此成为有价值的历史资料。这些照片分期拍摄于1918年、1921—1923年、1929—1930年及1935年。他热爱拍摄，到哪儿都带着相机，我们感谢他不仅拍了园林，还拍了中国建筑，尤其是北京的建筑，以及石窟中的雕像等等。

作为北宋（960—1127年）山水画大师之一，郭熙不需要过多介绍了。他是神宗（1067—1085年在位）最喜爱的画家，他的画作装饰着皇宫的大殿。其现存的最著名的画作是台北故宫博物院收藏的《早春图》，而由大都会美术博物馆（纽约）收藏的一幅手卷《树色平远图》（图2）同样是其重要的作品，并且与本文的讨论更为相关。这是一件短卷，最近冯良冰在其著作《灵验山水：论北宋宫廷绘画的权威性》以之为重点展开了深入研究。[3]

这件卷轴呈现了一幅河景的平远视图，开端是河上薄雾，两个渔人在舟中，远处是两个旅人牵一头驴消失在雾蒙蒙的山水中。继续展卷，有一座亭子为丛树和石头包围。亭子可看作园林的标志（将一处山水变为一座园林），或是山水中的观赏点。卷中人群走向亭子，证实了上述的后一个解读。其中一个仆人身背裹琴的长包，其他人带着野餐的物件。他们在亭中可以一边发现美景，一边享用食物聆听音乐。

文本《林泉高致》将我们引入郭熙的画论。[4] 郭熙强调山水画可引人进入一段旅途，称之为“卧游”。郭熙不是详释这一观点的第一人，早在四世纪宗炳（375—443年）对此已经有了讨论，而郭熙深入发展了这一观点，以四种不同的模式区分“卧游”的行进速度。[5] 据此，一幅山水画或可行，或可望，或可游，或可居。[6] 手卷的形式，尤其是那些被清楚的构以旅行知识的画作，最适合神游。郭熙的《树色平远图》长仅104.4厘米，易被看作一处单独的场景，图中所绘似一次短途旅行。不仅如此，他经营位置、装饰细节，引导观者在其中消磨时光而非匆匆一览：观者被吸引加入亭子里的活动，在那里和画中人物一起享受闲适的一天。

再观赏一件长些的手卷，郭熙的理念会显得更清晰。北宋佚名画家的《溪山无尽图》把观者带入一段长途旅行（图3）。[7] 画作以河谷景色为开端，气氛朦胧又令人期待。一条小路引导着“旅行者”（即画内的观者）入山后。[8] 当小路再次出现时，旅行者已抵达一座院落（有草顶和竹篱的乡村住宅）的门前，房子后面是山中平地。穿过院门，旅行者会看到山冈上的一座亭子。画外的观赏者会看到亭前临河，有房子在峭壁上；而旅行者看不到这些。小路拓宽，旅行者来到一处石矶（promontory），另一座亭子突出其上，下有船坞。一船正驶离，有人在送别。其他人聚于石矶上，有人身背长形包裹，也许是类似《树色平远图》中的琴包。旅行者可以跳上船，远观景致，沿河继续他的行程，也可以徒步原路返回，在随山蜿蜒的小路上选择右转。

小路伸向画面深处，消失在山中，旅行者此时身处路口，他需要决定是想随小路一同隐入他想象中的旅途，还是在路口左转，看看画家后面的安排。小路变宽，接入跨于溪上的小桥，溪水始自山里，汇入更宽的水路。小桥之后的溪岸上是一组房子，站在桥上，他会看到他之前去过的亭子。过桥，小路转入岸边巨石，退出视线，随后仅在汇入河流的宽一点的小溪旁出现。小路架于桩上，在山腰蜿蜒，旅行者无法过溪，只能漫步进入山中。可见此水不浅，人无法涉水而过，否则画家会画上涉水的人以示意。当旅行者穿行在山间，画面外的观赏者可以从远处欣赏山水风景。

如果旅行者决定跳过险峻的陆上行程而选择乘船的话，他将会更快抵达下一处聚落。小船自水上来，到达一处有亭子的石矶并于此靠岸。上岸后，旅途将在巨石和树丛后延续，到达茅屋环绕的一处开敞地。然后，稍远处恰可见一扇门的上端，正在探寻陆路的旅行者将从这座门进入村庄。村庄是休整之处，也是两种交通方式的转换之处。村子被遮住一部分，处于山脉的环抱之中。小路伸出村庄，跨过一条小溪：桥上有一人骑马、一人步行，正向旅行者走来。不过，细心的观察者在此又会看到小路的另一个方向，小路自村庄广场绕到巨石后面，引导观察者深入雾气笼罩的群山之中。

主要水道旁的道路却是一条更为易行的路，在这条路上可以与另一位孤独的旅人同行——也许可以同他一起在路旁那间外悬酒旗的简朴的酒馆歇脚。这路将会把旅行者引至一处寺院，寺中塔楼显于山巅之后。如果不是这条路而去寻找其他的路，目光仍被引入群山深处。对于画外的观赏者而言，庞大的山体在此处主导视野。潺潺流入主河的小溪中也许有一处水很浅，因为当大路再次出现时，岸边多了一些赶路的行人。观赏者的眼睛不再追踪大路的走向，而是发现了蜿蜒的岸线环抱中挤入的两只渔船。其中一只船上站了一位渔人，另一只船则空空的。我们的“旅行者”可以请渔人将他渡过水面，然而，望遍山峦与山谷，也看不到前路。此时应原路折返，或就地打鱼。

这件手卷以郭熙的“可行、可望、可游、可居”分类当如何观之？所有的种类(category)都是身处山水之中来体验；不同的种类指向不同的时间节奏，从较快的行走到在一处停滞。展卷之初，观赏者在聚焦细节之前会注意到开头画面的主要景色（如一处河谷）。观赏者在画内寻找山水征程的入口的同时，也在画外欣赏山水的呈现。画面《溪山无尽图》的迷人之处在于画家提供了不止一种观赏画卷的节奏。这里有所谓的主要线路，即顺着画幅下沿的河流前进，此处乘船是“捷径”。这条路与郭熙的“可行”相符。另一种较慢的展卷方式则是跟随伸入画面内部的分叉的小路。这些岔路是画中人的行进路线的一部分，画外的观赏者可以跟随这些小路，想象自己会如何在山、谷之中漫步。在赏卷的过程中，“旅行者”和画外的观者才是提供生动的线索以供神游的人。

在我的书《作为视觉的叙事的山水体验：李成—燕文贵传统中的北宋山水手卷》（2002 年）中，我聚焦于传为燕文贵（约 967—1044 年）画作的一件手卷。[9]《江山楼观图》为日本大阪市立美术馆藏。我在书中的主要观点之一是，即使一件手卷似乎并无故事可言，当我们观赏它时，我们的思维在观赏的过程中就会开始构建一个故事。大量细节将拖慢赏画的过程，这就是画家建立叙事节奏的手段，也是郭熙的分类的基础。他的有关绘制山水的要诀的讨论强调清晰与理性。一件成功的山水画应实现其主要目的，即协助观者神游其中。神游的需求来自年老、体病、休闲和冥想。这种对思维力量的信仰将我们导向中国的“理”的概念。北宋理学程颐（1033—1107 年）发掘了“理”与梦、记忆等等的关联：

> 心所感通者，只是理也，知天下事有即有，无即无，无古今前后。至如梦寐皆无形，只是有此理。若言涉于形声之类，则是气也。
>
> 天地之间，有者只是有。譬之人之知识闻见，经历数十年，一日念之，了然胸中，这一个道理在那里放著来。[10]

“理”决定了物质和对象的构建方式，可称之为万物的内在组织模式。“心所感通者，只是理也”。这一态度表明，山水画必须追随并表现同实景山水相同的“理”。如果偏离这个原则，画作必将失败，无法供人卧游。

研究和描述手卷的适当方式时逐帧展开，而非整卷摊开，那么，分析园林的适当方式应是漫步其中，而非对照一张总平面。

如前所述，在《中国的园林》的开头，喜龙仁提出了观赏山水长卷和游赏园林之间的密切关联。[11] 他强调一座园林不可从一个固定视点来观看；中国园林包含或多或少分离的碎片，这些碎片以有机的方式逐一相接，最终形成一幅统一的画面。在引论这里，他提倡将步履闲适的游园作为分析园林结构的基础。但是在之后分析园林个案时，喜龙仁并没有提及他游园的线路。实际上，诸如有关网师园的讨论仅占一段篇幅。[12] 苏州网师园是一座小园。此园引人流连、缓步、偶尔驻足欣赏景致和细部。园中挤满了建筑物，小路和连廊蜿蜒串联各部分。这是一座名副其实的迷宫，于我而言其缓慢展现，是完美的园林。景物环池布置：若环池游览，池塘时近时远，时而消失不见。

然而，游园旅程始于沿中轴规整布置相接的三座厅堂。游览者只有穿过这些厅堂，真正的园林迷宫才会开启。在三座厅堂之后，游人如果转入左边的小门，就会到达池岸上的亭子。作为园林旅程的真正的起点，此处宜停留并接收眼前展开的景致。这是步入山水之处。游人将立即注意到一簇荷花和正对面的“月到风来亭”。照片上只能看到其一角（图4）。[13] 亭左有廊连接至“濯缨水阁”，这是一处建于水上的平台，也是一座石洞的入口。[14] 若看向亭右，旅行者会注意到一座曲桥连接亭子。从他站的位置来看，这是最明显的一条路了。若要抵达小桥，他必往右转，走过环池建筑的平台，在穿过一处月门。立于桥上，旅行者向右可见“看松读画轩”，向左可见对岸“濯缨水阁”，在他走向亭子之前可欣赏水阁的角度随他的移动而变化。这“月到风来亭”是一处绝佳的歇脚处，可赏环池景致。此时园林入口在池对岸，乔木、灌木和石头以其后建筑物的白粉墙底为底，形成一幅令人愉悦的图画。如果恰逢中秋夜，其上可见满月，正合亭之名。片刻之后，旅行者沿着铺砌的步道走向“濯缨水阁”，即一开始入园时注意到的水阁。在这里，游人会再次回看园林入口（图5–图6），穿过门和建筑，他会来到曲桥。向左一瞥可见亭子的侧面（图7）。

在他的右边，一处石洞的入口引他进入，出口则是一座比“濯缨水阁”稍大的建筑。建筑之后他发现有几座小亭子，其中最小的叫“琴室”。这里已是尽端，也是一处令人愉悦的流连之所，折返可回到池边的主亭。再次穿过主亭，旅行者看到左边的一个入口，而进入一个较大开敞空间。这个空间为白粉墙界定，墙则为石头和植物的组合提供纸一般的白色背景，形成似灰色笔墨加少量绿色的水墨山水画的一处特写细部（图8）一类的从亭子越过池面看向入口时看到的景致。一座亭样的构筑物被置于门廊对面的长墙下。亭中有一块大石，座椅从两侧包围大石，此处既可做展示的平台，也可供休息（图9）。同那些石头和植物（的组合）一样，亭子的红色与灰色的石头一起，白墙为底形成一幅令人愉悦的图画。这块石头是画面的焦点，它是一颗太湖石：太湖石是名石的一种，曾被一些皇帝狂热的搜集，是许多园林中最值钱的宝物。徽宗（1100—1125 年在位）是最疯狂的收藏者之一，他在《祥龙石图》中绘制了一块太湖石。[15] “殿春移”在院子一头，值得一探，旅行者从边门出，来到“看松读书轩”——之前在桥上他曾瞥到此处。其门通往池塘；实际上这一整面都可以打开，接入景色和光线。至此，旅程大致在环池进行：小路沿池岸伸展，时而短暂地分叉至看不到池塘的地方，随后又回到池岸。在旅程开始时，旅行者意图抵达“月到风来亭”而穿过一些建筑物，现在他回到了这些建筑物中。他再次进入月门，发现此处还有一整片区域，可以早先游线的反方向进行探索。至“梯云室”，旅程似乎到了终点，这里暗示陆上旅程的最高点。这座园林设计由亭子、厅堂、院落多个空间单元细致构建，这些单元以类似迷宫的方式逐一相接，最终旅行者失去方向感和时间感。园林仅以“旅行者”（内部观者）的视点被观看。如果有人沿路拍照并将之拼成一幅全景图，会是什么样？

在一座中国园林中漫步时，大量的细部引人注意：有漏窗朝向不可进入的空间和空地（图 10），提示着前面有待探索的空间（图 11）。坐下来时可注视一处微缩山水（图 12），或注视脚下的景致（图 13）。由月门（图 14）或漏窗框起的景致如同画中画，这是一种再现的手法，常见于中国画。[16] 画中绘制屏风从而为画作增加含义。多数画屏没有其他含义：它们就是一件有时尚装饰的家具。然而，在大量的画中，屏中画像犹如一个新世界或是另一个维度的入口。后一种可举例如刘贯道（活跃于 1279—1300 年）《消夏图》，还有明人摩《重屏会棋图》。[17] 通过这些画作中的屏风图像，我们可在脑海中进入另一时空。窗中框定的山水（图 10）通往我们所在空间之外的景致，这也是一处不可致的景致。花窗的雕镂工艺增强了（对面的）图画世界的品质。月门可比作山洞的入口，作为景框它有了更多被阐释的可能性，因此它成为两个世界的边界：真实世界和想象世界。

在有关中国园林的西方文献中寻找有关网师园的描述，（不难发现）这些作者似乎极少在描述中采用（上述）漫步的方法，这也许是因为（这种方法）让描述变得复杂。上文我的描述中略去了许多细节和潜在的分支，此外，旅行者不断地回望，尤其会环视池塘，这让描述游园体验、尝试变得更具挑战。较常见的方式是使用通行术语讨论园林，依据拟定主题如水体、石头、诗歌、建筑等来排布章节。这样一来，如同在喜龙仁的《中国的园林》一书中有关网师园的材料被分散于若干章节中。在《一座中国的庭院：大都会美术博物馆中的明轩》中，网师园作为一个整体，得到了更全面的对待：读者有一张园林平面，结合平面有围绕园林不同角落的描述。[18]

陈从周（1918—2000年）于1984年提出园林可分作两类：

1. "静观"园林，其主要例证是网师园：他进而解释"静观"为"园中予游者多驻足的观赏点"。
2. "动观"园林，其主要例证是拙政园，即"有较长的游览线"。[19]

自陈从周的两种园林分类出发，我们可以发展出一套手卷和园林间的类比。我们是否可以修改陈氏的表述，说，可行、可望的山水画对应"动观"园林/山水？这样一来，可游、可居的山水画对应"静观"园林/山水？我们同样可以认为，像网师园这类园林是可游、可居的，像拙政园是可行、可望的。

有关绘画再现的类比不限于手卷，姜斐德和方闻在有关明轩的讨论中称"从中国园林的任一观赏点看去，比如在走廊中移动，景致如手卷一般展开。"[20] 实际上，从固定的观赏点看到的画面也可类比作画外观者欣赏的手卷，这就回到了我之前提出的问题：（如果有人沿路拍照并将之拼成一幅全景图，会是什么样？）例如，可将图3、图4、图5中的照片拼起来，得到一幅坐倚濯缨阁栏杆的全景图。我想说，视野并非如上所述或陈从周说的那样固定不变。即使身体是静止的，眼睛会继续游走，头部会继续转动。[21] 此前我认为在园林体验中不存在外部观者，然而更难准确地说是“旅行者”在园中以游览时在内部观者和外部观者的视角之间转换。“旅行者”在漫步移动时是处于内部观者的视角，而在驻足或是坐下环顾时则处于外部观者的视角。

程正揆（1604—1676 年）的手卷《江山卧游图》（150 号，1662 年作）为有关园林和手卷之间的关系讨论提供了一个有趣的对立点（图 15）。展开画卷，第一幅画面表现了一组建筑物妥当地安置在一处山冈后的山水中。冈上生有四棵虬树，冈后有河蜿蜒远去，冈左有桥跨河，于是我们知道“旅行者”有路可循。河对岸的房屋——仅屋顶依稀可见——则被掩于巨石和群山之间的谷地。从此，视野展开，溪流对面的山崖跌落瀑布，而小路向左蜿蜒至消失不见。这一刻有轻微的迷失方向，而继续展卷，山后有阡陌纵横现于谷间。在考虑向哪走时，“旅行者”突然遭遇一块巨大的太湖石立于山水间，几乎挡住了他的去路。这块石头来得蹊跷，旅行者和画外观者都停下来，边打量它边思考这奇遇。就尺度而言，此处还有其他不寻常。前景处，初看以为是大山的山顶长着云杉。然而，就整幅画面的尺度而言，那里却是大山的缩小版：这些更像是盆景中的石块间种植云杉苗。尽管这块太湖石（的出现）于两者（“旅行者”和画外观者）而言都有些莫名其妙，而这些不同寻常则对画外观者而言更为震撼。从“旅行者”的视角来看，这些盆景般的山水可看作远处的真山水。而在深入研究尺度的转换和惊奇的构成之前，应先完成画中的旅程。

至此，尚无任何人迹，仅房屋可以暗示山水的可居。旅行者走过太湖石，沿山崖左转（旅行者也可以选择另一条消失于山后的小路），他赶上了两位散步的学者。小路下行，有桥跨河。旅行者看到了上游河岸的房屋：而并无道路通往那里。为旅行者设置路径从前景向上进入群山。画外观者看不到这条路径了，他开始探索山间跻立的房屋。一座亭子的一角屋檐隐现于崖顶之后。其上，正有小路蜿蜒回归视野，小船浮于河上，有人向“旅行者”走来。之后，旅行者绕过一座陡峭而无法企及的山峰。

抵达山的另一边，“旅行者”立于一处石矶上，看见崖顶一座房子突出于水面上。崖下生有一棵粗糙又扭曲的松树。此处的风景由一片宽阔的水面主导，远山笼罩于雾中。两位学者立于此岸。其中一人向上指着飞鸟群——有可能是鹅。它们在左侧的河岸上，兀立一块太湖石，这块石头没有之前那块大。此后的风景愈发荒僻，“旅行者”需要一叶小舟去往近处的沙渚或是远处的群山。前景仅可见窄窄的一线水岸，而这却是目前唯一能向前走的山路。两位学者在稍远的岸边讨论着什么。此时，对岸的群山显得近了：一条小溪潺潺而下，上有小桥，又有小路。旅行者走过两位学者，一瞥小溪，转过巨石，抵达一处长桥，加入一队三人行。溪流显得湍急起来。过桥后，大山变得难以穿行。画外观者得以一览大山全貌，山体从画幅底端一直撑满到顶端。继续展卷，画外观者发觉旅程到这里就断开了，此后的风景对“旅行者”而言变得无法进入，他在山水中的路径被大山的崖壁阻断。此时，就余下的景物而言，这里有着与第一块太湖石周围类似的尺度转换。画幅底部的景物尺寸小，但是远景，同样的，直到卷末，山与水间的正常的尺度才会回归。

程正揆将此画命名为“卧游图”，而已知他曾将大量画作均作此名。他将这些“卧游图”赠送给他在北京新成立的满族政府的同僚们，本文所讨论的这一幅编号为150，[22] 这是一幅山水画，却有大量的细节，使人（观画而）如同漫步园中。画中的太湖石是最明显的园林的标志，画中尺度的转换也有着类似的园林的效果。此外，画中三两成群的人物，或谈天或赏景，其神态似闲适。园林与山水之间的关系，正如“旅行者”在网师园中的漫步，假山由石块垒成，带有真山的特点，旅行者需要解读自盆景向假山过渡的尺度转换。将程正揆的画解读作想象中的园林，这与李惠仪在“明末清初的园林和插画”中讨论的问题有密切的关联。[23] 她的研究表明，“有关想象的园林的写作潮流始于晚明”，这些作品带有梦境和幻想的魅力。[24] 这恰与程正揆的人生经历相合，他生于晚明，目睹了王朝的崩塌，后选择效忠清朝。李惠仪研究的大量园林仅存于纸上，有着诸如“乌有园”和“意园”这样的名字。[25] 这些文字呈现了在想象中的园林中如梦般穿行的旅程。这是一种坚持幻想 / 虚构而非真实 / 具体的写作趋势，反映了横跨十七世纪的变幻而纷乱的历史环境，这使得学者们——以及其他人——拒绝唯物主义，而再想象中寻求庇护。

正如我希望在这篇论文中表达的，园林与手卷之间的关系要比喜龙仁或其他人认为的复杂。这种比较并非全无益处；然而我们应察觉此三种艺术形式之间的微差，以及这种微差的多样性之广。我们也可以将外部观者的视角看作画家或园林设计师的视角。而手卷的欣赏者或园林的游览者则在两种视角，即创作者的视角和“旅行者”的视角之间摆动。当然，在“旅行者”的叙述之外，就山水和园林而言还有更多的东西。笔墨创造 / 描述的变化形成了画中以及园林中叙述的另一面。表面的图案和其所象征的物体（树、巨石、崖等等）之间有相互影响。然而这已超出了本文讨论范围。

在（本文的）结论部分，我们来看稍早的山水画中的园林，唐寅（1470—1523 年）作短卷《桐下喝茶图卷》，[26] 此画易被看作一幅单帧图像；然而，既然是件手卷，一开始，就有手卷的形式。展卷之初你并不知道卷的长度。你最先看到的（除了右下角的朱印外）是一座小桥跨于溪上，邀你过桥。与山同时，你细品小桥及其附近石头的画法。向上游看去，你注意到有男童在溪边洗碗。继续展卷，你会看到一位儒者手持茶杯坐在榻上，他对面有僧人坐在竹椅上啜茶。有仆童立于梧桐树下，刚从竹炉上煮好了茶。虽然他们看起来像是在短途旅行，实际上他们有可能是坐在园林中想象自己身处山水间。之后继续展卷，你会发现画幅在芭蕉丛之后骤然结束。这就是在邀请赏画者（回头）和他们一起饮一杯茶。

1

Figure 1 : Sirén's *Gardens of China* (1949) and *China and Gardens of Europe Eighteenth Century* (1990). Photo by Minna Törmä

图 1 : 喜龙仁的《中国的园林》(1949 年)和《中国与十八世纪的欧洲园林》(1990 年)。米娜 · 托玛拍摄

2

3

Figure 2 : Guo Xi, *Old Trees, Level Distance*, ca. 1080, handscroll, ink and colour on silk, image 32.4 × 104.8 cm. The Metropolitan Museum of Art. Public-domain image. https://www.metmuseum.org/art/collection/search/39668 (accessed 18.05.2021)
图 2：郭熙《树色平远图》卷，约 1080 年，手卷，绢本水墨设色，画幅 35.6 厘米 × 104.4 厘米。大都会艺术博物馆，公共领域图片。网址：https://www.metmuseum.org/art/collection/search/39668（2021 年 05 月 18 日访问）

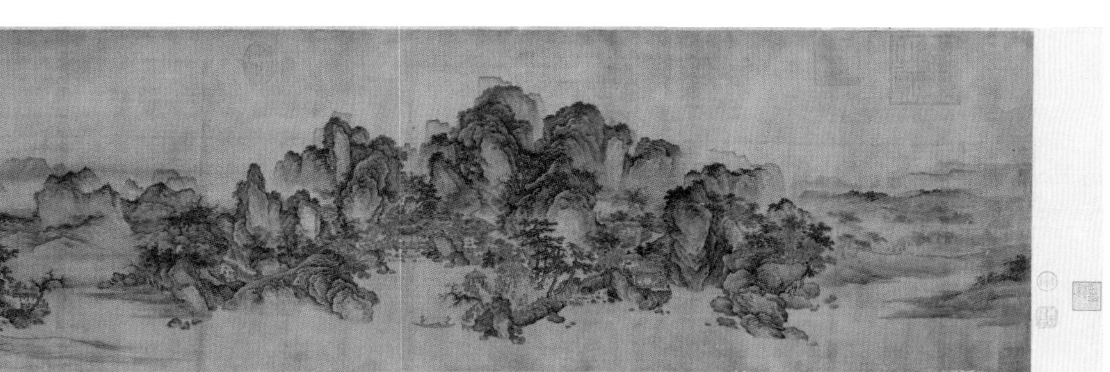

Figure 3 : Unknown Northern Song Artist, *Streams and Mountains without End*, 1100–1150, handscroll, ink and slight colour on silk, image 35.1 × 213 cm. Cleveland Museum of Art. Public-domain image. https://www.clevelandart.org/art/1953.126 (accessed 18.05.2021)
图 3：北宋佚名《溪山无尽图》卷，1100—1150 年，手卷，绢本水墨设色，画幅 35.1 厘米 ×213 厘米。克利夫兰艺术博物馆，公共领域图片。网址：https://www.clevelandart.org/art/1953.126（2021 年 05 月 18 日访问）

4

5

Figure 4 : Wangshiyuan: View over the pond towards "Washing Cap-Strings" pavilion. Photo by Minna Törmä
图 4 : 网师园：越过池塘看濯缨阁。米娜・托玛拍摄

Figure 5 : Wangshiyuan. Photo by Minna Törmä
图 5 : 网师园。米娜・托玛拍摄

6

7

Figure 6 : Wangshiyuan. Photo by Minna Törmä
图 6 : 网师园。米娜・托玛拍摄

Figure 7 : Wangshiyuan: Pavilion by the pond. Photo by Minna Törmä
图 7 : 网师园：池边的亭子。米娜・托玛拍摄

8

Figure 8 : Wangshiyuan: Arrangement of plants and rocks. Photo by Minna Törmä
图 8 : 网师园：植物与石头的布局。米娜・托玛拍摄

Figure 9 : Wangshiyuan: Pavilion for the rock. Photo by Minna Törmä
图 9 : 网师园：赏石的亭子。米娜・托玛拍摄

Figure 10 : Liuyuan: Lattice window. Photo by Minna Törmä
图 10 : 留园：漏窗。米娜・托玛拍摄

9

10

11

12

Figure 11 : Wangshiyuan: View through a window. Photo by Minna Törmä
图 11 : 网师园：透过窗子看到的景致。米娜・托玛拍摄

Figure 12 : Liuyuan: A tray landscape. Photo by Minna Törmä
图 12 : 留园：盆景。米娜・托玛拍摄

13

14

Figure 13 : Liuyuan: Decorated walkway. Photo by Minna Törmä
图 13 : 留园：精美的长廊。米娜・托玛拍摄

Figure 14 : Wangshiyuan: Moon door. Photo by Minna Törmä
图 14 : 网师园：洞门。米娜・托玛拍摄

江山卧遊圖 其一百五十
此圖作于辛丑十一月
予自金陵歸澤舟中
隨筆至二月十六日抵
家以人事置閣至除
夕檢點始竟計紙
四幅約長一丈六尺
青溪道人

15

Figure 15 : Cheng Zhengkui, *Dream Journey among Rivers and Mountains*, Number 150, 1662, handscroll, ink and colour on paper, overall including mount 36.8 × 921.7 cm. Los Angeles County Museum of Art. Public-domain image. https://collections.lacma.org/node/240953 (accessed 18.11.2018)

图 15：程正揆《江山卧游图》卷，1662 年，手卷，纸本水墨设色，整幅 36.8 厘米 × 921.7 厘米。洛杉矶郡艺术博物馆，公共领域图片。网址：https://collections.lacma.org/node/240953（2021 年 05 月 18 日访问）

Jingye: Crafting Scenes for the Chinese Garden

Antoine GOURNAY

In the Chinese tradition, gardens are constantly compared to painting. The art of composing gardens has often been compared to that of landscape painting (*shanshui hua*). Historically, these two arts have been simultaneously practiced and appreciated in scholarly circles. But despite the numerous links and mutual influences, which indubitably unite them, one cannot fail to observe the existence of significant differences between them, both in terms of the means they use and the ends they aim for.

Certainly gardens, like painting, offer us things to see, contemplate and comment on: they produce a show or spectacle.[1] But they are not built for this purpose alone. They are also designed as spaces for living, where it is possible to stay more or less long term, whether you are an occasional visitor or a permanent resident: they also produce a dwelling. The art of gardens aims at simulposition taneously fulfilling these two functions and reconciling them effectively.[2] The distinction between spectacle and dwelling is useful for analysis, but in reality they are inextricably intertwined. For example, the same architectural structure, such as a corridor or a pavilion, can play a role in forming part of the dwelling, because it allows one to move and settle comfortably, sheltered from the rain or the sun and away from others, and also in that of the spectacle, because it adorns the part of the garden where it is built and from it offers a pleasant view to contemplate (Figures 1–2). We will focus here on clarifying what it is which makes the spectacle in the garden specific, and which devices are involved technically in crafting it.

Although he claims to be a professional painter, Ji Cheng (1582–ca. 1642) focuses in his *Yuanye* on architectural features and on rockery-design, which were his speciality as a professional garden designer. Flower-lovers such as Chen Haozi (1612–after 1688) in his *Huajing* comment mainly on flower cultivation techniques. Later commentators usually describe the various elements by classifying them into categories (rocks, water, plants, buildings, etc.) that are examined separately and in turn, because different knowledge is needed for each type and their design and study concerns specialists in various fields.[3] However, the various elements used in the production of the spectacle, far from being simply juxtaposed or mixed randomly, are indissociably linked and organized within a coherent and constructed whole, intended to produce certain effects. More than the presence of certain particular ingredients, it is the way they are composed with each other, which is the specificity of the spectacle of the Chinese garden: it marries the living and the inert, the ephemeral and the permanent, the fixed and the moving. To better understand how it works, it is therefore necessary to study how all these elements come together and for what purpose.

From this point of view, the garden spectacle differs from that of painting in at least three ways. Firstly, it is not only visual but addresses all the senses of the viewer. Secondly, it is set out not only in space, but also unfolds over time. It is constantly changing, by itself, because some of its elements like plants or animals are born, grow and die, or are renewed cyclically. The passage of the seasons is marked by the presence of vegetal elements, and the abundance of places in the open air allows one to take notice of the weather and the passing time. The result is that one never sees the same garden twice. Finally, it unfolds in three dimensions and can be traversed physically and not only mentally, as when viewing a painting.

Far from presenting a continuous stream of impressions or a random kaleidoscope, the various elements of the spectacle are composed to form "views" or "scenes" distinct from each other and each having a particular interest. The word generally used in Chinese to refer to them, at least since the end of the Ming period (1369–1644), it seems, is that of *jing*, difficult to translate because it refers to both the thing to see and the awareness by the viewer of what is there, to be seen, worthy of being noticed and appreciated. Like its English counterparts, it is polysemous, and the meaning given to it has constantly varied over time or according to the context.[4] We will use here the word scene to designate these units, distinct from each other, that make up the spectacle and we will first examine how they are technically produced.

The habit of distinguishing places or corners of the garden according to the spectacle they present, the emotions they arouse, the manner in which one can settle there, and the activities that it is possible to practice there is long-standing. It dates back at least to the Tang (618–907), when the painter and poet Wang Wei (701–761) retired to his Wangchuan villa, close to the capital Chang'an, to engage in seated chan meditation and walk, play music, write and paint. He received his chosen friend Pei Di (v. 714–?) and composed with him twenty pairs of quatrains celebrating that number of distinct places of the garden, which inspired their inspiration and the communion of their feelings. Each of these places is indeed characterized by a particular atmosphere or charm, corresponds to a distinct state of mind and lends itself to different occupations. Wang Wei also represented them in painting, juxtaposed in a long panoramic scroll, between mountains and water. This work is now lost, but remains known through the many copies and reinterpretations it gave rise to. The garden itself has disappeared, but has remained for centuries an object of reverie and a major reference in the world of scholars.[5]

The spectacle is not just about producing sensations and impressions. At the same time, it evokes words and images to express them, which makes it possible to share them. It stimulates the imagination and inspires other creations, poetic, pictorial and musical. The scenes of the show are often given names and commented on. Titles and poems individualize and characterize the scenes, and suggest in a generally brief and allusive form the manner in which the spectator should consider them. They are often calligraphed and inscribed within the scenes themselves in the garden. The fact of distinguishing within the spectacle a set of scenes *jing* has gradually been systematised and practised on different scales, both inside small private gardens and on much larger sites, such as the West Lake in Hangzhou, which traditionally thas ten famous views (Xihu shijing). Often the repertoire of scenes enumerated within the same garden, constitutes a kind of programme intended to be discovered and travelled through, like an album of paintings accompanied by a parallel series of poems. The most developed programmes are those of the great imperial gardens in the Qing era: thirty-six views at Bishu shanzhuang, forty scenes at Yuanmingyuan.

As the architect William Chambers (1723–1796) remarked: "The usual method of distributing gardens in China, is to contrive a great variety of scenes, to be seen from certain points of view; at which are placed seats or buildings, adapted to the different purposes of mental or sensual enjoyments. The perfection of their gardens consists in the number and diversity of these scenes; and in the artful combination of their parts; which they endeavour to dispose in such a manner, as not only separately to appear to the best advantage, but also to unite in forming an elegant and striking whole."[6] These scenes can be composed using the most diverse elements. These are mostly visual compositions (as seems to indicate the etymology of character *jing*); however, many of them are not only visual, but aim at producing impressions on all the senses of the spectator. There are cases where the main interest of a scene lies in the olfactory or auditory impression that it produces.

In the Qinghuiyuan, at Shunde near Canton, the fragrance emanating from lotus leaves in the summer, when large drops of rain splash onto them, as in slow motion and with a dull sound, is part of a set of sensations for the person standing in the shelter of the kiosk at the edge of the pond, that are at once visual, olfactory, sonorous and even tactile, if we take into account the humid atmosphere in which the spectator is immersed and the clammy heat surrounding them. The water and the plants enable the garden to be filled with noises that the spectator identifies and appreciates as "natural", even if the devices that give rise to them are themselves entirely artificial: murmurs of the streams, lapping of the water features, rustling foliage, cracking of the bamboo. In addition, plants, rocks and ponds shelter all kinds of animals giving rise to sound by their movements and their cries. Birds, frogs or cicadas populate the gardens with their songs: it is the latter which constitutes the major element of the sound spectacle, characteristic of the garden, in which its inhabitants are constantly immersed. To all these sounds of natural origin is often added human music, sung or instrumental, for which the garden traditionally provides the desired setting for practice and appreciation.[7]

The scenes are not necessarily fixed, but evolve over time. The season, the time, the weather at the moment when the spectator discovers them come into play in their appreciation. They can include moving or intermittent elements: the wind blowing in the foliage, the quality of the light at a given moment, the appearance of the moon, the movements of animals or the water: one of the twenty scenes of the Jichangyuan, in Wuxi, is based around an artificial torrent that flows in several cascades between two walls of rock, creating pleasant sounds. They can thus constitute a permanent or ephemeral spectacle, which occurs only occasionally or cyclically: this is also the case of the famous scene formed each spring by willows planted at the edge of Thin West Lake, in Yangzhou. One can perceive here a clear taste for the ephemeral and cyclic nature of the spectacle of the garden, which follows the rhythm of the seasons and thus of the universe: to capture it is, for the spectator, a means of being a witness of this rhythm and participating in it oneself. Gardeners thus play upon the transformational properties of some of the elements that go into making the show. Certain effects, delayed in time, such as the blooming of trees or the change of colour of their leaves, because they are expected, can be foreseen in advance. Therefore, the spectacle is partly programmable.[8]

The focus of interest may be one single element. It may be, for example, a tree, which is considered particularly venerable because of its shape or age. The Yuhuayuan, the imperial garden located to the north of the Forbidden City, on the main axis of the palace, has a beautiful centuries-old sophora (*huaishu, Sophora Japonica* L. alias *Styphnolobium japonicum* [L.] Schott), which covers, from its branches supported by a forest of crutches, a surface of several tens of meters squares, thus constituting a sort of covered shelter. It is still alive and its flowering is looked forward to each year as an important event. It is then time to come to contemplate the scene: the word *jing* is used to designate it by its contemporary commentators.[9] The Liuyuan, in Suzhou, features a famous Taihu rock with fantastic shapes, particularly appreciated because they strike the viewer's imagination. Its profile is slim; it seems to rise to the sky in a spiral movement. Through some of its holes, one can see light; others are nothing but black and mysterious crevices. Despite its astonishing twisted figure, he stands, and stands alone, imposingly, in the middle of a courtyard. An undeniable impression of balance and unity emerges from the whole. This rock has been named Guanyun feng ("Peak Crowned by Clouds"), a name that invites the viewer to see the image of a mountainous summit rising to the sky, and even to imagine it surrounded by clouds.

But gardeners do not always have such rare or exceptional specimens to hand. If they are not available, they ingeniously combine several elements the combination of which combination must arouse the interest of the viewer. Some scenes are composed of two elements that contrast but balance with each other, interact and complement each other, and mutually enhance each other. We observe all kinds of combinations: it can be a rockery (*jiashan*) falling perpendicularly into a pond in which it is reflected, a kiosk perched on the side of a hillside or isolated in the middle of a pool, or the combination of bamboo with rocks. They are brought together at the same time for the contrasts they form when they are juxtaposed, and the equilibrium that is established between them. Thus, the bamboo associated with rocks contrast by their greenness, their flexibility, and their apparent fragility with the greyness, the hardness and permanence of the stone with which they are confronted. But they are not only close together for the physical sensations produced by their juxtaposition. When the gardener composes bamboos and rocks, he does not mix only the hard and the soft, the green and the grey, but he combines elements immediately identifiable by the spectator as being bamboos and rocks. These elements refer to one's perception and conception of them, which have already been expressed, in the form of words and images, in poetry and painting, and also in other gardens, by other similar compositions, which pre-exist. When he composes them, the gardener therefore refers to a pre-existing model that serves as a reference. In addition to the rocks and bamboos, there is a symbolic meaning: they embody each of the different virtues, which are those of the virtuous man (*junzi*). Bamboo, with its constantly growing suppleness and vigour, is in fact a complement to the longevity and solidity represented by the rocks. This association of bamboos and rocks is based on the traditional representation in Chinese culture of these two elements, which includes impressions, shapes, words and symbols.

The informed spectator, who contemplates them in his turn, does not only see combinations of colours and forms; he identifies the elements he has before him as bamboos and rocks, he knows that the way they are arranged refers to this known model, he recognizes symbols. The content of this representation is both perceived and conceived. Other scenes bring together three elements that balance each other. We thus observe, forming a trio, pines, bamboos and Chinese plum (*mei, Prunus mume* Siebold and Zucc.). The garden creates here the same type of image that is frequently represented in painting, but in three dimensions and using the real elements as the material. If one is familiar with the scholar's culture, seeing this scene, one can name the famous trilogy, frequent in painting since at least the time of the Song (960–1279): that of the "Three Friends of the Winter" (*Suihan sanyou*): In the cold season, when all other plants are dormant, bamboo and pine remain green, and the plum is the first to bloom. It is traditional to admire them as a symbol of solidarity in resistance to adversity[10]. In this association, which has become traditional, we find here again, in addition to the production of physical sensations, the image, the discourse and the symbol at the same time. The trilogy has a name that has become almost proverbial and the actors appear almost like characters or actors of the scene presented.

Other, more complex scenes, involve a greater number of elements, as can be seen in the Zhuozhengyuan, in Suzhou, with a scene in a small inner courtyard called Haitang chunwu ("Low wall and crab-apples in spring"). This title, calligraphed on a small relief scroll-shaped panel at the top of the wall forming the back of the scene, indicates that the scene is supposed to be contemplated in spring, at the moment when the two small trees on either side are flowering. Once the season has passed, the sign and the fixed elements of the device continue to evoke it and even if it is no longer visible, the imagination of the spectator is thus stimulated at any time by the reminiscences that the reading of the title elicits (Figures 3–4).

The scene proposed for the contemplation of the spectator is often organized around a main or central element which dominates it and which is at the same time showcased by other elements that surround it. Thus, a building may be the centre of interest or the star element: this is the case of most of the scenes listed and named by the emperors Kangxi and Qianlong in their gardens: they almost all coincide with a creation constructed in a valley or on an island. But it can be just as well a group of plants, a lake, a hill or an artificial mountain. The way in which these various elements are composed responds to aesthetic preferences which are, of course, variable, but some of which have been defined and codified by scholars who have, over the centuries, developed a particular taste, which can be found in painting as in poetry and which influenced all the gardens. It is impossible to summarize it in a few words, but we can note some characteristic features. Instead of seeking to multiply the effects, the educated gardeners prefer, for maximum impact, to restrict them in number: it is therefore an aesthetic of the stripping away, restraint, simplicity. We can also observe, in the way that elements are arranged spatially a taste for asymmetry, for the sinuous rather than the rectilinear, for what seems random but which is in fact totally mastered and balanced. Of course, the reverse tendencies could also be observed in the Chinese tradition of gardens, especially in the imperial gardens, where there would be a preference for the grandiose, symmetry and regular order. But even in these, the aesthetics of the scholars had a certain influence: the Emperor Qianlong tried to imitate in his gardens the spirit of what he had seen in the private gardens of scholars during his travels in the South.

In any case, the way in which the ingredients of the spectacle are composed to form scenes refers to established custom: certain associations have thus become habitual and are expected by the spectator, who is supposed to have knowledge of these customs to fully appreciate them. They constitute a sort of repertoire of traditional motifs, which garden designers constantly come back and refer to, proposing something analogous to musical variations on a given theme. They are produced in the garden in a form that most often follows rules which are implicit, but similar to those that have been explicitly codified for painting, for example in the famous encyclopaedia entitled *Jieziyuan huazhuan*. This explains the frequent connections that are made between painting and the art of gardens. They partly share the same imagery, which is also found in poetry. This is also why Ji Cheng, the author of the *Yuanye*, constantly refers in his book to these two arts, and himself gives a poetic and allusive form to his explanations.

So far, we have grouped under the name of spectacle all that, which is offered for contemplation by the spectator present in the garden. But it is also possible to dissociate from the spectacle devices that we will call "spectation", the aim of which is to show the scenes to the viewer and to show them in a calculated and pre-conceived manner.[11] What is then crafted is not the thing to be viewed but the device that makes one look at it differently.

The limits given to a scene may very well exist only in the view of the spectator who selects elements from what they see, and associates them: they are often invited to do so by the internal coherence of the composition that is proposed to their eyes. However, certain elements can be used to limit the visual field spatially, while being themselves part of the spectacle. They give autonomy to the scene in relation to what surrounds it, isolating it from the rest of the spectacle, which helps to highlight it and attract the viewer's attention. In the largest gardens, where there is a lot of space, the different scenes can be enclosed by hills or artificial slopes which thus form isolated valleys, each containing a different scene to contemplate: a process commonly used in imperial gardens in the Qing Dynasty. This was observed by the Jesuit missionary brother Jean-Denis Attiret (1702–1768) describing what he called the "valleys" of the Yuanmingyuan.[12] Scenes can also be circumscribed by the boundaries of a clearing, or occupy an island in the middle of a body of water. The ponds and lakes thus play an important role in the enhancement of the elements that surround them. As already noted in the eighteenth century by the architect William Chambers (1723–1796): "They compare a clear lake, in a calm sunny day, to a rich piece of painting, upon which the circumambient objects are represented in the highest perfection; and say, it is an aperture in the world, through which you see another world, another sun, and other skies."[13]

But when you are inside an urban garden of smaller dimensions, the view is rarely so extensive; generally, it is limited, to a fairly close radius, by the raised rocks, curtains of trees or bamboos, and especially by the walls and the various buildings, which enclose the scene to contemplate and play the role of insulators.

In Jiangnan gardens, it is common for a wall to be used as a backdrop for the presentation of a scene. There are several examples in the small courtyards of Shizi lin: the fact that they are covered with a plain white coating has often afforded the comparison to the sheet of paper used by the painter. But here the wall is clearly separated from the element which it contributes to highlight: it modifies by its presence the way in which the spectator perceives what is placed in front of him. These white walls also serve as screens on which the shadows of various elements of the spectacle are projected. In his book, Ji Cheng particularly recommends the patterns thus created by the foliage of the scholar-trees projected on the walls[14]: the large, compound, bipinnate leaves of this tree lend themselves to effect to this game, because they project onto the wall countless small moving shadows that animate the surface.

On a smaller scale, a stone planter or a simple pot can act as insulators for a scene they contain. Some scenes are thus presented separately or isolated in vases, trays or planters, which can be fixed or mobile. In Chinese this is usually called a "view in a dish" or a "potted scene" *penjing*. The material from which the container is made, its colour, its shape, are important because they influence the way in which its content is perceived by the spectator. Gardeners create plant and rock compositions, sometimes rocks only in gravel or water. Planters or pots placed on various supports present their contents at a calculated height so that they are best seen by the viewer. Small *penjing* have the advantage of being able to be moved and exposed to different parts of the garden, both inside the rooms and outside, for limited periods, for example at the time of flowering of the plant. But what constitutes the profound originality of the art of *penjing* is the deformation of some plants, and the miniaturization that is sometimes the subject of the scene presented. For some trees, this is actually done by means of dwarfing techniques, which not only artificially reduce the size of a living plant by controlling its growth, but also give it reduction, the appearance it would have or should have had it been allowed to reach its natural size, or if it were much older, or if it were growing on the side of a high windswept mountain. Here again, one tries to make the tree conform to a reference: to the idea that one has of the aspect that it could have, under certain conditions, in its natural size. This idea is partly based on trees observed in reality, whether in the wild or cultured, and it is also influenced by the multiple representations already existing of such trees, in the form of verbal descriptions or images. The miniaturized trees of *penjing* are both real trees and the image made of other real or imagined trees.

But it is also possible to resort, to miniaturize a scene, to illusionism. The most common way is to create an optical illusion by simple juxtaposition: the size to which a tree has been reduced

leads the viewer to consider the pebbles that surround it as being on the same scale, therefore seeing an impressive cluster of rocks. Or the scale of the scene to be seen is sometimes more explicitly indicated by introducing miniature constructions into the composition, such as bridges or pagodas, paths and stairs or even small characters. Their very small scale encourages the viewer to look at their surroundings as much larger than in reality. Without being able to penetrate physically into these *penjing*, the spectator is invited to perambulate there mentally, by imagining himself reduced to the scale that was indicated to him.

However, whether there is miniaturization or not, the tray or the pot used to contain these *penjing* play an essential role in the way in which the viewer considers its content: they make these scenes of the worlds closed, independent. In the Chinese tradition, that which is enclosed preserves its nature intact: life takes place in a closed circuit, regenerates itself. Because of their isolation, these potted scenes acquire the fullness and inviolability of the closed world. The power of life manifested there, especially in the venerable form, as if contorted by age, of a rock or a tree where it seems concentrated, exerts a beneficial influence on what surrounds them and particularly on one who contemplates them.[15]

Other processes consist of intervening not on the object of the spectacle, but on the spectator themself, by positioning them at a calculated distance from the scene to contemplate. For example, the artificial mountains on which he climbs give him access to lookouts less open to the outside, from which he can contemplate from high and far other parts of the garden. The balustrades are a means of inviting the spectator to come and lean over them, and thus draw his attention to a scene to contemplate. Terraces are specially designed to allow the viewer to contemplate a particular scene from a selected height and angle while maintaining a calculated distance between them and the scene to be viewed.

The name given to some of these terraces sometimes suggests that they serve as observatories to contemplate the moon. In addition to offering an unobstructed view of the sky, the fact that they are almost always built on the edge of a pond or lake makes it possible to see the reflection of the moon, and to enjoy the particular brilliance of the water surface when the moon is out. Such processes, specific to the Chinese garden, are called "borrowed scenes" (*jiejing*).[16] The aim is to highlight various objects, which can be outside of, or independent from the garden, and have not been made for the purpose of the scene, and to integrate them in the spectacle. The expression is ambiguous, because it designates both the borrowed scene, which is part of the spectacle, and the device that allows one to perceive it, which is that of spectation. Whatever the nature of the element borrowed, what interests us here, from the point of view of the spectation, is what is crafted to make it perceptible to the spectator. In the Yuyuan, in Shanghai, it is a two-floored pavilion *lou* that allows one to contemplate the distant river beyond the wall of enclosures. When you take, as in the Zhuozhengyuan, the view of the pagoda of the Northern temple that pre-exists the garden, it is the whole garden that becomes a device to enjoy this view.

In other cases, it is a question of making the spectator aware of natural phenomena that cannot be directly controlled, but which are known to occur at one time or another. Mist, wind, rain or snow are thus integrated into the spectacle and "staged," by means of devices that make it possible to highlight them: the viewer is thus able to better appreciate them, to enjoy a "front-row view" peacefully. Awnings and covered galleries allow the spectator to enjoy the elements while remaining sheltered. The pavilions built at the edge of the water allow one to see and hear from close up the drops of rain that hit the surface or the wind that draws ripples water. Just like water, plants can play a role in these devices of spectation: this is the case of bamboos that quiver and crack in the wind in a particular way, recognizable for the connoisseur. It is also that of the lotus leaves, on the glossy surface of which the drops of rain drum and bounce with a dull sound. The Geyuan, in Yangzhou, has wall with a series of orifices, intended to serve as a musical instrument to the winter wind.

But the most original process used in the Chinese garden is perhaps that of the vertical framing of the scene to contemplate. This type of device makes it possible to enclose at a distance the object or scene to be viewed in a frame through which one sees it. The location and shape of certain openings are thus calculated so as to frame a scene that is contemplated through it. When inside the main pavilion of the Yuyin shanfang in Panyu, the viewer sees an erected rock just in front of one of the windows, which is framed and highlighted for him (Figure 5). The distance at which the rock stands, the format and the orientation of it are obviously calculated so that, detached from its real and immediate environment, the rock imposes itself on the spectator. This process makes it possible to attract and focus the viewer's attention on the thing to see, by placing it at a distance and as in a place beyond: it is visible, but it is part of a different world. The shape given to the outline of the opening is important for the way the framed scene is perceived by the viewer. The frame plays both the role of a dissimulator, by the limits it imposes, and a revealer of the thing to see, by the particular point of view it proposes.

The various types of openings, doors or windows, placed in the partitions create a number of frames usable to present selected scenes to the spectator. Two types of openings are used to frame and present scenes: on the one hand, the doors and the "empty windows" (*kongchuang*) which form only an outline which marks out the scene to contemplate, and are entirely clear inside; on the other hand, the "flowery windows" or "ornate windows" (*huachuang*), the openings of which are lined with various elements that partially obstruct the view (Figure 6). In some cases, they include in their centre a kind of viewfinder through which, when approaching, one can look at the scene beyond (Figure 7). Empty doors and windows have contours of various shapes: there are rectangular openings, pentagonal, hexagonal, octagonal, round, others that evoke the silhouette of an object: flower corolla (quadrilobed plum flower or begonia with oval petals) vases, bottles, gourds, fans (Figure 8). The Liuyuan, in Suzhou, offer a rich variety, which in itself is one of the elements of the garden spectacle. But all these forms are not only used to enrich the show with additional decorative elements; they help to introduce the scenes they frame by giving them a particular value or meaning.[17]

The particular shape given to the frame signals to the viewer the interest of the scene that one can see through it. The long, narrow shape given to some of the vertical or horizontal frames is reminiscent of the size of the scroll paintings (Figure 9). Other windows have the shape of a fan, one of the formats frequently used in painting. The view that we see through is therefore supposed to evoke, through the format in which it is inscribed, the paintings with which we adorn these familiar accessories. The Zhuozhengyuan has a fan–shaped pavilion whose walls are pierced with windows of the same shape, framing views of the water feature and the islands in front. Other openings have the outline of a vase or gourd, traditional Taoist symbols of the heavenly world where it is difficult to penetrate by a narrow bottleneck. The scene contemplated through the opening of the window or door thus appears as a different and desirable world. In his book, Ji Cheng recommends that "wooden walls should have many window–openings, so that one can secretly enjoy looking through them into different worlds, as if in a magic flask."[18]

Doors are also used to frame a scene. This role is mainly devolved to those practiced in walls of masonry, which connect different parts of the garden. What differentiates them from the windows is of course that they can be crossed, and thus seem to invite the spectator to do it by presenting to him, framed, a part of the different world to which they give access. The many circular openings are called "grotto doors" (*dongmen*), which again refers to the Taoist mythology, or "moon doors" (*yuemen*) (Figure 10). Their form evokes that of the moon, so important in gardens, and makes them, without a doubt, elements of the spectacle. But these doors are at the same time devices of "spectation", since they allow the spectator to perceive, framed by them, the scene which is beyond. The inscriptions that name or poetically comment on the scenes reinforce these framing devices, as much as intellectually in the mind of the spectator. Indeed, the very arrangement of the signs around the doors itself forms a frame: the front plate (*bian*) bearing the title of the scene is placed above the opening, and the two vertical panels inscribed with parallel sentences (*duilian*) which comment on it, are suspended on each side.

Other devices take into account the fact that the spectator moves around inside the garden: they construct the way the spectator is led to discover the show during his travels and thus come under spectation that one could qualify as kinetic or motional. They act on the viewer as an incentive to browse and explore the garden. When several frames accumulate in a row between the spectator and the thing to be seen, it appears even more distant and like a mise en abyme. The revelation, thus made to the spectator, that there are several successive spaces beyond the one in which he finds himself, is also an encouragement or a stimulus to discover them. The spectator is clearly invited, when it comes to doors, to cross them one by one to get closer (Figure 11). Thus these framing devices not only manage the way to contemplate isolated scenes, but also allow the journey of the viewer through the garden to begin. The fragmentation of the spectacle into scenes does not reduce it to a series of vignettes. The scenes can be related to each other, so as to form what could be called a "programme". At each new stage of the journey, a scene that the spectator had first seen from a distance and which had attracted him may constitute, once reached, a new starting point. Thus in the Liuyuan, Suzhou, when the viewer enters one of the courtyards of the entrance, the existence of the next courtyard is revealed to him by openwork windows, which let him see a part of it in advance. The fact that he is presented with a path that seems to lead there, in the form of a door to cross, encourages him to move forward to get there. Another process consists of creating contrasts between scenes contemplated successively. For example, in the Wangshiyuan in Suzhou, from the "Pavilion of the Small Hill with an Osmanthus Grove" (Xiaoshan conggui xuan), the viewer is presented with various scenes, including a cliff of rocks through an "ornate window" on the north side of the pavilion, with a vertiginous precipice; but if he turns round, he is immetdiately confronted, through the main opening on the south side, with a much more peaceful scene, combining fragrant osmanthus and bamboos. The way in which the different scenes answer each other spatially is thus calculated: they reinforce each other's effects. "Small Hill with an Osmanthus Grove"is a common title given to similar scenes in literati gardens which refers to a famous verse from the *Songs of the South* (*Chuci*), in the *Summons for a Recluse* (Zhao Yinshi): "Osmanthus grows thick in the mountain's recesses," which alludes to the idea of reclusion in the mountains.[19]

The distinction that has just been made between fixed and kinetic spectation is useful for this analysis. But it would be arbitrary to divide gardens into two opposing types: ambulatory and sedentary gardens. In fact, all gardens use more or less of both at the same time, and often the two types of spectation I have distinguished complement each other and reinforce each other. It is clear, however, that the larger gardens, such as the Zhuozhengyuan in Suzhou, tend to use kinetic spectation techniques that encourage the viewer to explore them to every corner. After a while, the unfamiliar visitor in the garden does not know clearly how he got to where he is, where he actually is, or how he can come out of the garden. At the same time the spectacle, highlighted by the spectation, captivates his senses and gives him pleasure. But even in the smallest gardens, such as the Liangyuan, in Foshan, space also seems, thanks to the combined devices of spectation, to expand indefinitely and allow the spectator to enter a kind of dream world (Figure 12). In fact, we cannot go around these gardens in just one visit: these are just made to offer new experiences over a lifetime.

景冶：中国园林中的造景技艺

顾乃安

（顾凯、叶聪、严佳 译）

在中国传统中，园林总与绘画类比，造园艺术则常同山水画艺术进行比较。历史上，这两种艺术在文人圈中同时得到实践和欣赏。但除去那些无疑能将二者统一起来的诸多关联和相互影响，我们也一定要注意，无论是在手段采用上还是在目标指向上，二者都存在重大差异。

不可否认，园林像绘画一样为我们提供了观看、沉思和评论的对象：这里成为一种展演（show）或景观（spectacle）。[1]但这些并不是建造园林的唯一目的。园林也被设计为大体能够长期停留的生活空间，不管对于偶然的访客还是常年的住户，这里还是一处居所。园林艺术旨在同时满足这两项功能，并使二者有效和谐。[2]景观和居所之间的差异有助于开展分析，然而在现实中，两者间有着错综复杂的联系。举例而言，同样的建筑物，如廊或亭，在居所的构成中有其作用，能使人舒适地走动和栖身，并免于风淋日晒之苦；而对于景观而言，此类建筑使所在的园林部分生色，在此也可获得供人凝望深思的愉悦景观（图 1–图 2）。本文将重点阐释园林中的景观显得特别的原因，以及对此的精心营建所涉及的方法。

尽管自称为专业画家，计成（1582—约 1642 年）在他的著作《园冶》中重点关注了建筑特色和假山设计——这些才是他作为专业造园师的专长。花卉爱好者，如《花镜》的作者陈淏子（1612—晚于1688年），主要讨论花卉栽培技术。后世的评论家们通常将园林中的各种要素进行分类（石、水、植物、建筑等等），各类别被依次分开探讨，因为每种类型需要不同的知识，对其研究和设计需要不同领域的专家。[3]然而，在景观形成中所使用的这些要素并非简单并置或随意混合，而是在一个连贯构筑的整体中得到密切关联和组织，意在制造某些效果。正是其相互构成的方式，而非某些特定成分的呈现,形成了中国园林景观的特异性——生机与无机、瞬息与永恒、稳定与运动，都融于一体。因而为了能更好地理解其中机制，有必要去研究所有这些要素是为何、且如何组织在一起的。

从这一角度来看，园林景观至少在三个方面与绘画不同。首先，园林不仅是视觉的，而且作用于游览者的各种感官。其次，园林不仅是空间的布置，而且随时间而展开。由于如植物和动物这样的要素在生老荣枯，或周期更新，园林自身在持续改变。植物要素的状态呈现着季节更替，充足的露天空间也使人注意到天气变换与时间流逝。其结果是，从没有人能够两次见到同样的园林。最后，园林以三维形式展开，能够被亲身游览，而不是像绘画一样只能进行神游观赏。

远非呈现出连续的印象流或随机的万花筒，组成景观的各种元素形成了相互不同、各有特色的“景象”（view）和“场景”（scene）。对此，至少从明代（1368—1644年）末期开始，中文里通常用“景”这个词去描述。这个词难以翻译，因为它既指所观之物，也指观赏者对存在何物、看到何物、何者又值得被注意和欣赏的意识。就像它在英文中多样的对应词一样，随时间和语境的不同，“景”的含义在不断变化。[4]这里将使用“场景”（scene）一词来表述这些组成景观的互异单元，并且首先检查它们是如何被具体制造出来的。

根据对景观的呈现、对情感的引发、可供游憩的方式、可进行的活动来对园林中的场所进行划分，这一习惯由来已久。这至少可以追溯至唐代（618—907 年），诗人画家王维（701—761 年）退隐到临近长安的辋川别业，在此坐禅、行游、操琴、写作和绘画。他和挚友裴迪（714 年—）相互唱和了二十首律诗，赞美园林中二十个使他们灵感激发和情感共鸣的场所。这每个场所确实都有其特定的氛围或魅力，对应各种特别的心境，适合进行不同的活动。王维也用绘画来呈现这些场所，将其并置于全景长卷之上、山水景象之中。虽然这幅作品久已佚失，但仍通过诸多摹本及仿作而闻名。辋川别业本身已经不存，但多个世纪以来一直作为怀想的对象和学界的重要参考。[5]

景观不仅使人产生感受和印象，它同时还能够引发人们用语词和图像进行表达，进而使得对其共享成为可能。它启迪想象力，并激发其他创作，如诗歌、绘画和音乐。场景经常被命名和讨论。标题和诗歌使场景个性化、特征化，并往往以一种简短而用典的形式，暗示观者应该对待场景的方法。它们也经常被书写并镌刻在园林的场景之中。在景观中区分系列之“景”的行为逐渐变得系统化，在不同尺度上都有实践，无论是小型的私家园林还是更大的区域，如杭州西湖传统上有著名十景。通常，在同一园林中所列出的各种场景，构成了一种旨在被探索和游赏的节目安排，就像一本附有相应系列题诗的画册。最成熟的是在清代皇家园林之中，如避暑山庄三十六景、圆明园四十景。

就像建筑师威廉·钱伯斯（William Chambers，1723—1796 年）评论的那样：“在中国，布置园林的通常方式是设计各式各样能从某些视点观察到的场景，在这些视点上设置座位或建筑，以满足精神或感官享受的不同需要。园林的完美在于场景的数量与多样性及各部分间的艺术整合；在这样的处理方式下，各部分不仅各自以最佳方式呈现，而且合成为一个优雅而醒目的整体。”[6] 这些场景可以由迥异的要素组成。其中主要是视觉构成要素（正如“景”的字源所体现），但许多也不只是视觉的，还意在创造观者的其他感官印象。在一些例子中，场景的主旨就在于对嗅觉和听觉印象的形成。

在广东顺德的清晖园中，夏日荷叶散发着清香，大滴雨点儿打在叶片上，发出沉闷的声响，继而缓缓滑落，再加上所沉浸于内的黏潮空气的氛围，这些都是池边亭中的观者能够体验到的视觉、嗅觉、听觉、甚至触觉的一部分。流水和植物使得园林充满着能被识别并欣赏的“自然”声响，尽管产生这些声音的设施完全是人造的：溪流潺潺、叠水哗哗、树叶瑟瑟、竹裂声声。此外，植物、山石和池塘为各种动物提供了庇护所，其活动和鸣叫也产生各种声效，鸟、蛙、蝉的鸣声使园林充满生机：这正组成了声景的主要要素，构成园林的特色，令人沉浸其中而流连忘返。在这些自然声响中，经常混杂有人为的音乐，或歌唱、或奏乐，传统上，园林为其提供了表演和欣赏的理想环境。[7]

这些场景并非一成不变，而是随着时间而发展。观者游览时的季节、时间、天气都会对体验造成影响。其中包含移动的或间歇的要素，如穿过枝叶的风、某一时刻的光影、月亮的出现、动物或水的活动。无锡寄畅园二十景中的其中一景，就是以人工涌泉为基础，水流在两侧石壁间一再跌落，发出悦耳的声音。由此，它们可以构成或永久或短暂的景观，偶尔或周期性地呈现。同样如此的还有扬州瘦西湖畔的垂柳，每年春天就会成为著名的风景。人们能够清楚地感受到，园林中的景物遵从季相更替和宇宙变化的节律而呈现周期性和短暂性的特点。对于观者来说，捕捉到景物的这种特性是见证自然规律的一种方式，而自身也参与到其中。造园家们发挥了某些要素的可变特性，从而更好地营造出场景。某些效果会随延时到来，如人们所期待的树木繁花盛开或叶色变化，都可以被提前预测。因此，景观在某种程度上来说是可设计的。[8]

场景的焦点可以只是单个要素，比如一棵树，因为树形特殊或者树龄古老而备受推崇。在紫禁城主轴线北部的御花园里，就有一棵美丽的数百年古老的槐树。靠许多支架撑住枝桠，它的树冠覆盖了数十平方米的地面而形成荫蔽。该树依然生气十足，每年的花开都会被期待为一场盛会。这时人们开始深思这个场景，时评家会用“景”字来描述。[9]苏州留园有一座造型奇特的著名太湖石，因其能够激发观者的想象而为人们热衷欣赏。这块太湖石的外形纤细，似以螺旋的动态上升，直指云天。石头的某些孔洞可以透光，其他却只是暗黑而神秘的裂隙。除开令人惊异的扭转形态，此石孤独而庄重地立在庭院正中，一种无懈可击的平衡与统一的印象从整体中涌现。这块石头被命名为“冠云峰”，使观者联想到直插苍穹的山峰，以及环绕飘荡的云彩。

然而造园家们并非总能获得这样的珍奇之物。如果材料不尽如人意，他们会巧妙地将几种要素组合起来以引起游人的兴趣。有些场景由两种彼此对立而又平衡的要素组成，相互影响，相辅相成。我们观察到各种各样的组合：假山从水池中立起而又倒影于池面，亭子坐落于山麓或孤立于水中，或是竹子和岩石相配。这些要素并置时形成了对比，同时又能相互平衡。如竹与石之间，苍绿与烟灰、柔韧与坚硬、易朽与永恒形成了鲜明的对比。但它们间的紧密关系并非只来自因并置而产生的物质性感受。造园家布置竹与石，不仅混合了软与硬、绿与灰，而要让两者能被人们一眼识别出来，唤起人们对这些要素的感知和观念。在此之前，它们就以文字或图像的形式在诗歌和绘画中，以及其他园林中，以类似的构成得到表达。如此，当造园家创作此景时，就能引证已有的典范作为参考。竹与石还有另一重象征含义：它们身上体现着不同的君子美德。竹子具有不断生长的柔韧和活力，岩石则代表长寿与坚忍，二者构成互补。竹石的这种关联，正是基于二者在中国文化中的传统再现，如印象、形态、文字和符号。

熟悉此文化的观者，欣赏时不仅观察到色彩和形式的组合，而且识别出他面前的要素（如竹子和石头），并能够认出文化的符号、理解组合的原型。这种再现的内容，既受到觉察，也得到构想。此外还有三种要素相互平衡的场景，即以松、竹、梅形成“三人组”。绘画中也经常表现这一类形象，园林中则在三维空间中用实物来营造。对着这一场景，熟悉士大夫文化的人就能指出这一组合为著名的“岁寒三友”，这至少在宋代以来的绘画中就已频现：在寒冬腊月，其他植物全都沉寂的时候，竹和松仍然长青，而梅花傲雪凌霜率先开放。作为保持团结、对抗逆境的象征，“岁寒三友”作为传统而得到尊崇。[10] 可以再次发现，这种早已成为传统的意象组合，除了能够引发人的知觉感受外，还同时有着图像、诗文和符号象征。“岁寒三友”几乎家喻户晓，其中主体就如同舞台场景中的演员角色。

其他一些复杂的场景涉及更多的要素。例如在苏州的拙政园中，有一个名为“海棠春坞”的内向小院，这一题名以书法呈现在卷轴状的匾额上，作为小型浮雕置于白墙的上部、场景的背后，暗示着欣赏场景的时间应在春季，此时两侧的小株海棠会盛放花朵。当花季结束，虽然花开的盛景不再可见，但这一符号及固定的要素设置依然能够将此场景唤起，通过阅读题名而引起的回忆随时激发着观者的想象（图3–图4）。

为静观沉思而设计的场景，通常围绕着一个主导的或中心的要素而组织，同时也被周围其他要素所衬托。这时，建筑物就可能成为场景的中心，如康熙和乾隆皇帝所造园林中的题名景观大多如此，或坐落于山谷，或处于岛上。然而场景的重点也可以是植物组群、水池、丘岭或是假山。这些不同要素的组织方式来自审美的偏好，当然这些偏好是可变的，但其中一部分由士大夫所定义和确认，数百年来已发展出特定的品位，在绘画、诗歌中都有体现，也影响到园林。这种偏好很难用寥寥数语概括，但可以罗列出一些特征。训练有素的造园家们并不是去追求丰富多彩的效果，而是倾向用最少的数量来获得最佳的作用。因而这是一种关于剥离、克制和简练的美学。同样可以观察到，园林要素在空间上倾向于不对称布置，多使用曲线而非直线，布局看起来随意，却是相互平衡的精心安排。当然，相反的倾向在中国造园传统中同样存在，尤其是皇家园林中更强调富丽堂皇、对称和秩序井然。但即使在这里，士大夫美学也有着一定的影响，乾隆帝南巡之后，就在皇家园林中极力写仿他所见过的私家园林。

各种场景的要素组成方式，都要按照既定的惯例：某些关联为约定俗成，从而符合观者的预期，而观者也要对这些惯例有所了解，才能加以充分地欣赏。这些组合构成了传统主题的某种“曲库”，造园家对此不断地加以回顾和参考，创作出类似音乐上对于某特定主题的变奏曲。园林中的这些实践通常遵循一种暗含的形式规则，又类似于绘画中明白流传的画谱，比如著名的百科全书式的《芥子园画传》，这也可以解释绘画与园林艺术之间的密切联系。二者部分地共享相同的意象，这在诗歌中也可以找到。这也是为什么计成在《园冶》中不断提及这两种艺术，并以诗意和喻指的方式来行文的原因。

目前为止，我们将园林中供观者品赏的都归为“景观”。但是也有与景观设置有别的“观法”（spectation），其目的是以一种经过计划和构想的方式向观看者展示场景；[11] 这里被精心安排的不是被观景物本身，而是景物的布置方式，使人的观赏可以耳目一新。

对场景的限定可能是在观者的视野中自发形成的，人们通过眼中所看到构图的内在连贯性，从所见中提取要素并将它们关联起来。不过，某些要素在自身成为景观中一部分的同时，还可以用来限定视觉的空间场域。它们通过将场景与周边的景观其余部分既分离开，又产生联系，而使场景产生自主性，从而有助于突出该场景并吸引观者的注意。在大型园林中，空间辽阔，不同的场景可以被山丘或人工山坡所围合，从而形成包含着差异性场景的独立山谷，这是清代皇家园林中常用的手法。耶稣会传教士王致诚（1702—1768 年）观察到了这一点，并描述了圆明园中的所谓“山谷”。[12] 场景也可以被空地的边界所围绕，或者对于水体中的岛屿，池塘与湖泊包绕着这一景物并加以有力地强调。正如十八世纪的建筑师威廉·钱伯斯（1723—1796 年）所说：“在宁静的晴日，他们将一方清澈的湖面比作一幅丰富的画作，在此，水面围绕的景物得到最完美的再现；这是世界的一个窗口，通过它，你会看到另一个世界，有着另一个太阳及其天空。”[13]

但当你置身于尺度较小的城市园林中时，视野很少如此广阔。方寸之地通常被峰石、树幕、竹林，尤其是围墙与各式建筑所限定，它们起着屏障的作用，使场景被围合而得到品赏。

在江南园林中，以墙为背景来衬托景物是很常见的，如在狮子林的多个小庭院中就有几例。覆盖着白色涂层的墙体常与画家的用纸相提并论，但这里的墙与它所要烘托的要素是明显分离的，墙的存在改变了观者对眼前之物的感知方式。同时，这些白墙也起到了屏幕的作用，各种景观要素的阴影投射其上。计成在其书中特别推荐了槐叶之影投于墙上而形成的图案[14]：无数细碎而婆娑的树影映于墙上，生动有趣。

在较小的尺度上，一个石盆或小罐可以用来界定某个场景。因此，有些场景被分别呈现在花瓶、托盘或植盆中，可以固定，也可以移动。汉语中通常称此为“盆景”。容器的材料、颜色和形状都很重要，因为它们影响着观者对盆中之景的感知方式。匠师创造植物和岩石的组合，有时仅在砂或水中置石。放在各式支架上的花盆或罐子以计算好的高度呈现盆中之景，以呈现最好的观赏效果。小盆景的优势，是可以在有限的时段，比如植物开花之时，在园林的室内和室外不同地方移动并展示。但让盆景艺术具有深刻独创性的，是对某些植物的变形，以及有时呈现景观主题的微缩。对某些树来说，这实际上是通过矮化的技艺来实现的，不仅通过控制植物的生长来人为地缩小活体植物的尺寸，并进行修剪而使它呈现出自然的效果，或貌显苍老，或长于悬崖。人们试图使树符合某个模式，即它在一定条件下可能具有某种自然面貌的观念。这种想法，部分是基于在现实中观察到的野生或是栽培的树木，同时也受到已有对这类树木的口述或图像形式的各种再现的影响。盆景中的微缩树景，既是真实的树木，也是一种来自其他或真实、或想象的树木的意象。

但也可以依托于幻觉艺术来微缩场景。最常见的方法是通过简单的并置来创造一种视错觉：由缩小的一棵树，观者会觉得周围石块与之尺度相同，那么壮观的群石就得以显现。有时在构图中引入微缩建筑，如桥梁或宝塔、小径和台阶，甚至是小人，能更加明确地呈现所要的场景尺度。它们的极小尺度，使观赏者对眼前之境的感受要比现实之景大得多。虽然人们无法身入其境，但却能想象自己被缩小到同等尺度，进行精神上的游赏。

然而，无论是否进行微缩，用来承纳这些盆景的托盘或罐子在赏景方式中起着至关重要的作用：它们使场景世界封闭而独立。在中国传统中，自成一体使本性保持着完善，生命在循环往复中生生不息。正由于其独立，这些盆景在自我的世界中拥有了完整性和不可侵犯性。其中的主体山石或树木，以其饱经风霜而令人动容的姿态，呈现出生命的力量，并对其周遭、特别是品赏者产生积极影响。[15]

还有一些手法安排的不是景物而是观者本身，将观者定位于一个与所品赏的景物有着特定距离的位置。例如，攀登假山使人到达一个不向园外开敞的高处，于是可俯视或远眺园林中的其他区域。栏杆可以吸引观景者倚靠其上，引导其关注某个场景。平台往往是精心设计的，使观赏者以特定的高度和角度，并以特定的距离来品赏某个特定的场景。

有些平台的命名暗示着它们是用来赏月的观景台。它们几乎都建于池边或湖畔，因此在皓月当空之时，除了可以一览无余地观看天空之外，还可以欣赏月亮的倒影、享受水面的别样美景。这种为中国园林所特别关注的手法被称为“借景”，[16] 其目的是对一些外在或独立于园林的、并非特意用来造景的物象，将其突显并融入景观之中。这种表达是模糊的，因为它既指作为景观一部分的所借之景，又指作为观法的让人感知景物的方式。从观法的角度来看，不论所借元素的性质如何，吸引我们的是能让观者感知到匠心所在。上海豫园中有一座两层的楼阁，让人可以隔墙眺望远处的江水。

在拙政园，你可以远眺早于此园就已存在的北寺塔，于是整个园子就成为观赏此景的一个装置。

另外一些案例中的问题是，对那些时而会发生的、不可控的自然现象，如何得到观赏者的关注。通过一些手段，雾、风、雨、雪被突出地融合到景观中，如同舞台演出，于是观众能够更好地去加以欣赏，如“前排效果”一样自如享受。檐与廊使观者能够在荫蔽下观赏风景，水边的亭子让人可以近距离看到和听到打在水面上的雨和吹起涟漪的风。和水一样，植物也可以在这些观法的装置中发挥作用，比如竹子在风中会有特殊的颤动和声响，而得到行家的欣赏。荷叶也是如此，雨滴在荷叶光洁的表面上敲打，发出沉闷的声响。扬州个园有风音洞墙，意在成为一种让寒风演奏的乐器。

但在中国园林中最具创意的手法或许是框景，是将一定距离下的物体或场景，让人从一个框中加以品赏。因此需要考虑这些开口的位置和形状，以便框住要品赏的景色。在番禺的余荫山房的主亭内，观赏者能看到一座石峰立在一扇窗前，这扇窗就是用来框定并彰显这一峰石的（图5）。峰石所在的距离、形式和朝向显然是经过精心考虑的，使石头脱离其紧邻的现实环境而跳入观者的眼帘。这种手法能吸引并聚焦观者的注意力，看到位于一定距离远处的景物：它既可见，却又遥不可及，似在另一世界。开口的轮廓形状影响着观赏者感知框景的方式。景框有着双重的意义，既造成限制而隐藏起景物，又产生特定视点而揭示了景物。

隔断中各种类型的洞、门或窗，形成了大量的景框，把选定的场景呈现给观者。有两种类型的开口用来框住和呈现场景：一种是门和“空窗”，它们只形成一个轮廓以展示要品赏的景观，框内之景得以完整呈现；另一种是“花窗”，开口内有各式纹样，遮挡了部分视线（图6）。有时，花窗的中心会有一个取景口，人走近时可以通过此处看到其外之景（图7）。空门窗的轮廓有各异的形式，如长方形、五边形、六边形、八边形、圆形，还有一些呈现着物体的轮廓，如花冠（花瓣呈椭圆形的四瓣梅花或海棠）、花瓶、罐子、葫芦、扇子（图 8）。苏州留园中有着丰富种类的花窗，其本身就是园林景观的要素之一。但所有这些形式不仅是用来丰富展示的附加装饰，还有助于引出所框之景，并赋予特定的价值或意义。[17]

景框的特殊形状让观赏者更能透过景框而获得场景之趣。一些竖向或横向的狭长框形，让人联想到绘画中的卷轴（图 9）。有的窗户则呈扇子的形状，这也是绘画中经常使用的一种形式。因此，透过这种景框所欣赏到的景象，就让我们感到正是熟悉的画中景物。拙政园有一扇面亭，墙上又开了一个扇形窗洞，框住了亭前的水景与岛景。另一些开口的形状则呈花瓶或葫芦形，它们在道教传统中象征超凡的世界，通过狭窄瓶口艰难穿越而可到达，因此，透过这样的窗洞或门洞而品赏的景观就显得别有洞天、令人神往。计成便在书中写道："板壁常空，隐出别壶之天地"。[18]

门也可以用来框景。这样的门洞主要在砖石墙上，也把园林各部分连接起来。它们与窗的不同之处自然在于可被穿越，于是似乎在邀请游赏者前往一个部分已框景呈现的另一天地。许多圆形开口被称为"洞门"，指涉着道教的神话，又称"月门"（图 10）。这种形状无疑使人联想到园林景观的要素——月亮。同时，这些门洞也是"观法"的装置，它们使观者能感知到所框住的远景。对场景的命名或诗意评论的题刻，也在观者的观念中使这种框景作用得到加强。事实上，围绕门洞布置的题刻就形成了一个景框：题有景名的匾额置于门洞上方，而刻有评论此景的对联的两块竖板挂在两侧。

另一些设置考虑到了观者在园林中移动的事实，构建起引导观者在游赏中发现的方式，因此也有一些适于动观的观法。这些设置诱导观赏者游览并探索园林。当多个景框在观者和景物之间叠加，呈现出嵌套结构，更显遥远幽深。观者从而得到一种启示，即发现除自己所处的空间之外，还有几个连续的空间，这也鼓励或刺激着人们去进一步探索。当接近门洞时，观者无疑被吸引着去一一越过而接近这些空间（图11）。因此，这些框景装置不仅安排着各个离散景点的品赏方式，也开启了观者的穿越园林之旅。园林景观分散为多个场景，并未使其减弱成涣散的各种片段；场景之间可以相互关联起来，于是形成“游线”。在每段行程开启时，游客刚一抵达，就会一下子远远看到吸引他的景物，于是构成一个新的起点。如在苏州留园，当观者进到入口的一个庭院时，下一个院落即通过漏窗向他显现，让他提前看到部分内容。在他前面，有一个可穿越的门洞，引导一条隐约能通向那里的小径，吸引他向前走去那里。另一手法，是在接续的场景之间形成对比。例如苏州网师园，从“小山丛桂轩”中，观赏者看到各种场景，如透过北面花窗看到山石峭壁，但转过身来，便能立刻从南面主门看到更为幽静的桂竹相映场景。不同的场景通过这样在空间上的呼应，加强了彼此的效果。“小山丛桂”是文人园林中类似场景的常见称谓，指的是《楚辞》中《招隐士》的名句：“桂树丛生兮山之幽”，暗含隐于山中之意。[19]

上文对静观和动观的区分有助于分析，但若把园林分为动与静两种对立的类型就过于武断。事实上，所有的园林或多或少都会同时使用二者，而且两类观法往往相辅相成、相得益彰。不过对于大型园林，如苏州拙政园，倾向于采用动态观法技巧，鼓励观赏者去探索园中各个角落。不熟悉此园的游者很快便会失去清晰判断：如何到达这里，究竟身处何处，又要如何离开。与此同时，由观法所凸显的种种景观又让游客欲罢不能并心生愉悦。但即使是在小型园林中，如佛山梁园，也由于结合了观法的设置，空间似乎无限地扩大，让观者恍入梦境（图12）。事实上，这些园林是我们无法一次游遍的，即便一生反复来游，也会每次都有新的体验。

1

2

Figure 1 : Zhuozhengyuan corridor
图 1 : 拙政园走廊

Figure 2 : Zhuozhengyuan fan-shaped pavilion
图 2 : 拙政园扇面亭

3

4

Figure 3 : Haitang chunwu close-up view
图 3 : 海棠春坞近景

Figure 4 : Haitang chunwu enlarged view
图 4 : 海棠春坞中景

5 6

7

Figure 5 : Yuyin shanfang vertical frame window
图 5 : 余荫山房竖框窗

Figure 6 : Liuyuan corridor with ornate windows
图 6 : 留园走廊与花窗

Figure 7 : Zhuozhengyuan part view
图 7 : 拙政园局部景观

8

9

10

11

Figure 8 : Ouyuan corolla-shaped window
图 8 : 耦园花冠形空窗

Figure 9 : Zhuozhengyuan empty windows
图 9 : 拙政园空窗

Figure 10 : Liuyuan moon window
图 10 : 留园月洞门

Figure 11 : Liangyuan entrance
图 11 : 梁园入口

12

Figure 12 : Liangyuan bridge over reflected kiosk
图 12 : 梁园拱桥及倒影

“知夫画脉”与“如入岩谷”：清初寄畅园的山水改筑与十七世纪江南的“张氏之山”[1]

顾凯

1 引论

寄畅园历来为江南名园，自明代中期建园后便多有文人赞誉，晚明以来名声日隆，清康熙乾隆二帝的多次临幸，尤其是乾隆帝的大量诗作吟咏并在北京清漪园加以写仿，更增其名望；虽然历经沧桑，今日遗存的寄畅园仍然具有极高的造园艺术水准，许多专家推其为现存江南园林之首。[2] 寄畅园在历史上曾有过多次大改，最后一次重要改筑发生在清初，[3] 直接奠定了今日所见的山水格局面貌，对理解当代园景特点最为关键，因此对此次改筑值得加以深入认识。

康熙《无锡县志》有“寄畅园”条，是作为主撰人的当时寄畅园主秦松龄所作，其中有这样的描述：“园成，向之所推为名胜者一切遂废”，[4] 说明清初的改筑对园林的原先面貌有极大改变。然而对于究竟是怎样改筑的，一手的历史资料非常有限，缺乏直接描述的园记和直观的园图等材料，相比前一次晚明的秦燿改筑所遗留资料的丰富详细（如宋懋晋所绘园景图册、王穉登所作园记等），[5] 相差极大；因此当代的寄畅园历史演变研究中，对晚明营造的论述较详，而对清初转变的认识相对较弱。在已有的对寄畅园清初改筑的研究中，黄晓的成果相对丰富，进行了时间标定、观念梳理和特色总结，[6] 也对具体营造内容做了一定的推测，[7] 为进一步的研究打下了重要基础。此外，秦志豪对一些相关历史过程做了考证，是较新的研究进展。[8] 总体而言，现有研究对寄畅园清初改筑如何使旧日名胜“一切遂废”而脱胎换骨，还不够深入、清晰，甚至有论者在较近的研究中仍认为“张鉽修葺的寄畅园在景观构成上较秦耀时期没有太多变化”，[9] 显然与历史文献有相违之处。

对于直接史料相对欠缺的问题，我们注意到可以从相关历史认识中获得助益。康熙《无锡县志》提及了这次改筑的主持匠师：“云间张南垣琏，累石作层峦浚壑，宛然天开，尽变前人成法，以自名其家，数十年来，张氏之技重天下而无锡未之有也。至是以属琏从子名鉽者，俾毕其能事以为之。”[10] 这次改筑是明末清初最杰出的造园家张南垣委派其重要传人、其侄张鉽主持，这提醒我们对此次改筑的认识，不仅要关注营造活动事件本身，还要结合更宽广的园林史视野，尤其是进入十七世纪明末清初江南园林山水营造转变的具体历史情境来考察。

对张南垣及其所开创的以“张氏之山”闻名的造园叠山流派，已在学术界得到一定深入认识，[11] 而张南垣本人的叠山作品在今天“一体无存”，[12] 张南垣的传人张鉽所改筑而成的寄畅园山水，是今日江南所见“张氏之山”的唯一遗存。因此，通过“张氏之山”的研究，可以对清初寄畅园山水改筑做法得到更为深入的理解，也更进一步认识今日寄畅园山水的艺术成就；与此同时，结合现状及其他历史材料而对清初寄畅园山水改筑的深入考察，又可以反过来作为江南仅存例证而进一步认识“张氏之山”，从而深化和丰富对园林史的理解。

对于清初寄畅园山水改筑的研究，还将纳入与更多的其他时期（尤其是晚明与当代）园貌的关联比较来进行。根据历史文献及研究，从晚明秦燿之后至今，除了清初这次改筑，作为寄畅园核心主景的山水格局并未进行过其他重要变动，因此，清初改筑前的山水状况主要是基于对秦燿所筑寄畅园的认识，而改筑之后则主要借助各种文献（包括清中期的图像）以及今日所见的遗存面貌。尽管历史上确实屡有破坏和修缮，建筑、植物的变动相对频繁，但就基本山水面貌而言，寄畅园的今日所见仍然为清初改筑所奠定，因而对此次改筑成果的认识也可以很大程度上通过当代平面图及现状照片作大体了解，尽管会有不尽准确之处。

2 "张氏之山"：十七世纪江南造园转变的引领

曹汛指出："张南垣为我国首屈一指的造园叠山大师，他的造园叠山作品水平最高数量最多，古今中外没有一个人比得过。他开创一个时代，创新一个流派，对我国的造园叠山事业作出了极大的贡献，他的成就把我国的造园叠山艺术推到巅峰，对当时和后世造成了极大的影响。"[13] 晚明江南造园发生了深刻转变，张南垣正是最重要的推动者和代表人物。[14] 张南垣的影响力还不仅在晚明，入清后他仍多有作品，而且有子侄作为他的传人，可以说在十七世纪的明末清初都极具影响力。[15] 张南垣所开创的叠山流派，时人称之为"张氏之山"，如陆燕喆《张陶庵传》有："南垣先生擅一技，取山而假之，其假者，遍大江南北，有名公卿间，人见之，不问而知张氏之山"，[16] 可见风格的鲜明。

对于"张氏之山"特点的具体认识，可以从两方面来认识：一是视觉形态，二是游观体验。

在视觉形态方面，曹汛总结为四点：一是"以山水画意通之于造园叠山，有黄、王、倪、吴笔意，峰峦湍濑，曲折平远，巧夺化工"，二是"反对罗致奇峰异石，反对堆叠琐碎的假山雪洞，提倡陵阜陂陀、截溪断谷、疏花散石，随意布置"，三是"提倡土山和土石相间、土中带石的假山"，四是"综合考虑园林布局，有机安排山水与建筑及花木的配置"。[17] 其中第一点对画意营造的总结可谓纲领，后三点正是第一点的具体实现。在晚明得到确立的画意宗旨，正是造园转折的深层动因，[18] 而张南垣正是画意造园叠山的杰出引领者。吴伟业《张南垣传》描述时人对他的评价："华亭董宗伯玄宰、陈征君仲醇亟称之曰：江南诸山，土中戴石，黄一峰、吴仲圭常言之，此知夫画脉者也。"[19] 作为文化领袖的董其昌和陈继儒从其作品中看出了元代画家（黄公望、吴镇）的风格，并以"知夫画脉"为对他叠山成就的最高赞誉。迥然不同于以往对奇特、宏丽的大型石假山的崇尚，"张氏之山"以画意为宗旨，尤其崇尚元人笔意，追求简明之法、天然之趣、象外之韵。

"张氏之山"的特色还不止于画意形态，更有对游观体验的关注。以张南垣为引领的十七世纪江南造园的巨大变革之中，游人在园林山水中的动态游赏体验越发重要，加入了游观体验，由"景"上升到"境"，成为山水营造更为重要的目的。[20] 在后人对张南垣的评价中，这一方面也正是其中要点，如戴名世《张翁家传》中"君治园林有巧思……虽在尘嚣中，如入岩谷"，[21] 蓝瑛、谢彬纂辑《图绘宝鉴续纂》中有"张南垣……半亩之地经其点窜，犹居深谷"，[22] "如入岩谷""犹居深谷"之语，都表明"张氏之山"对身体浸润其中的进入式真山水体验的极大重视，并成为其标志性的成果。

这样的视觉形态与游观体验二者之间是有着密切内在关联的，都是画意造园追求的组成部分。画意造园不仅在于视觉形态效果，游观体验也是重要方面。笔者曾有论述，山水画意绝不仅仅意味着画面、构图的欣赏，还重在精神性的漫游；在山水绘画理论中，无论是画家的创作还是观者从画中所得，都不是静态的，而是需要"游"的存在。当画意确立成为园林的宗旨，对于园林的欣赏和营造也随之产生深层次的影响，从着重关注离散景点中的静观，到逐渐关注动态的游赏，行进过程中的体验成为重要欣赏内容。《园冶》中的"拟入画中行"成为画意追求下对空间性园林山水动态体验关注的最贴切形容。[23] 黄宗羲《张南垣传》中所谓"荆浩之自然，关仝之古淡，元章之变化，云林之萧疏，皆可身入其中也"，[24] 也正是对张南垣造园的"知夫画脉"视觉形式和"如入岩谷"境界体验二者密切关联的极好描述。

在理解了“张氏之山”的基本特点之后，我们就可以更有效地关注清初寄畅园的改筑究竟是如何进行的，又怎样使旧日名胜一切遂废而脱胎换骨；或者说，究竟改了些什么，又好在哪里。由于张南垣与张鉽的造园特色主要在于山水营造，“张氏之山”的说法也可见其擅长，而寄畅园也正是以山水之景闻名，这里主要就山水营造方面进行探讨。

在张鉽改筑完成后的诸多寄畅园诗文中，常见对园中两处山水景象作重点并列描述，如邵长蘅《惠山游秦对岩太史寄畅园六首》诗中有“扑帘苍翠逼，罅石洑流涓”，[25] 余怀《寄畅园闻歌记》中有“循长廊而观止水，倚峭壁而听响泉”，[26] 从中可见对园景有这样两大关注：其一是主景山池区，可以在池畔亭廊观“止水”池景及隔池扑面而来的“苍翠”山景；其二则是山中的涧泉，可以结合“罅石”“峭壁”而赏“洑流”“响泉”。对于清初寄畅园的山水改筑，也可以从这样两个重要位置区域来认识：一是作为主景的山池结合区，一是作为特色的山中谷涧区，这两方面也恰好与“张氏之山”的两大特色相对应。

3 “知夫画脉”：山池的画意经营

首先来看主池“锦汇漪”及相邻假山一带，这里作为全园主景，山与池的大体格局，奠定着全园中最基本的山水关系。

主景假山与水体的建设工程浩大，许多晚明著名园林在这方面耗费的财力和时间投入都是巨大的。[27] 然而清初的财力与明代盛时差距极大，寄畅园的全园改筑仅用一年多也就完成，[28] 如何能在短时间内，运用少量的人力、物力、财力代价进行改造，而能做到脱胎换骨般的效果提升？

这一神奇任务的完成，无疑来自改筑者的巧妙构思。总体而言，在改筑前后，假山部分和水池部分各自的位置、体量，及相对关系都没有作很大改动，然而组织的方式有了非常大的变化。

对于晚明秦燿改筑后的寄畅园面貌，由于当时存留的大量文字与图像资料，已经得到黄晓等学者的深入研究，并有平面复原图来加以较为清晰地呈现（图 1）。从这些材料和研究中可以看到，晚明寄畅园的山水主景区，总体而言有着较为复杂的营构，景点较多，建筑分量较重。在假山部分，有曲涧、悬淙亭、爽台等景点设置，并且在假山之上树立多处显著石峰加以欣赏（图 2），即屠隆所谓“峨然奇拔者为峰峦，窅然深靓者为岩洞”。[29] 在水池部分，则更与建筑密切联系，最显著的营造是贯连的长廊并穿越于水面之上，长廊在水中央设“知鱼槛”（图 3），成为整个水域的中心位置，能最佳欣赏两侧水景，类似现在杭州郭庄“两宜轩”的作用。知鱼槛及相连长廊，将池面大体划分为南北两个区域：在南区，在北侧的知鱼槛与西侧长廊所连的先月榭和霞蔚斋、东侧的蔷薇幕，围合出一片水面，从“先月榭”之名看到这里可欣赏初升之月的倒影；在北区，则以水中岛上的涵碧亭为核心（图 4），这里能欣赏西侧假山及涧水飞瀑（图 5），本身也能作为戏台，在其南侧的知鱼槛、西侧岸边廊中清籞斋等处加以观赏。

这样以较丰富的建筑营造来进入山水的安排，游赏者可以在较为舒适的行进与休憩中从容欣赏多样景致，整个园林也可以因建筑的精致营造而显得富丽堂皇。[30] 这种追求符合当时造园风气，与假山的“峰峦”“岩洞”一道，建筑的“华堂栉比，高阁云齐，飞梁若虹”，[31] 是当时寄畅园所为人赞叹效果的重要组成部分。

然而在十七世纪造园风气的变革中，这些曾受到赞赏的营造则显现出问题，既由假山的方式，更由山水中的大量建筑营造而带来：长廊及其连接的建筑不仅割裂了水面而使水景不够突出，更使山水相对分离而严重削弱山水关系，山水自然的整体感知仅存在如涧水飞瀑的局部，对山水景致的欣赏显得过于涣散而不够整合统一。总体而言，在更追求天然山水的雅朴简洁和整体画意的明末清初，这种分量过重的人工建筑及其造成的对自然感知的支离破碎，已严重不合时宜。

再来看清中叶的寄畅园图像（图6-图8）以及与之较为一致的当代寄畅园山水格局（图9）：假山已经不再有“峰峦”“岩洞”，而是以土山为主，点缀叠石，其上植物丰茂，似真山坡麓，正如清初顾景星对邵长蘅《惠山游秦对岩太史寄畅园六首》中的点评道：“垒石不作峰峦，而多陂陀漫衍之势”；[32] 池面空间旷朗，池面之上与山池之间不见任何建筑营构，仅在池面东侧形成亭廊，可以隔水观山，另于池面北部架设桥梁，以通两岸。

通过山池面貌的前后对比，大致可以看到清初对此改造的要点：一方面，对于假山，以典型的“张氏之山”进行改造，形成土山为主、错之以石的“平冈小阪，陵阜陂陁”，起伏有势，配以林木，辅以石径，宛若天然；并设亭于山的后部，以合画意。[33] 另一方面，对水池区域，变动最大，池中及池西与假山相接的一侧，所有建筑被去除，甚至涵碧亭所在的岛屿也消失了，仅在东侧形成亭廊，在池西岸则营造出山水相接（如鹤步滩）的自然效果。总体效果上，大片水面得以完整呈现，水景尤为突出，而山水关系上，不同于此前仅以涧水流瀑为山水的结合，山与水得以完整地密切配合。

这一改造确实形成了与此前差异极大的强烈景观效果，我们可以从池面横、纵两个方向来考察。

首先，在东西横向上，形成了鲜明的隔水看山的格局：位于池东岸的“知鱼槛”等亭廊，成为重要的山水景致观赏场所，游人在此对望池西山水（图10）。在此所观效果，最大限度地利用了位于惠山东麓的地理位置条件——从此处隔水西望，惠山位于假山之后，园中“多陂陀漫衍之势”的假山，不仅本身如真山冈阜，而且更似惠山的坡麓余脉，从后往前绵延而达水池西岸，使园中山水与园外真山有一气相连、一体延展之感，从而形成绝妙“借景”——从园中望去，园林山景顿时延伸至远方惠山之上，园林景象空间大大拓展，园景突破了用地的局限而真正与园外天地之景融为一体，这正是一个优秀园林所追求的崇高境界。

清中期著名文人袁枚《秦园》之诗的开篇即为“为高必有因，为园必有藉。美哉秦家园，竟把惠山借”，[34] 将此视为寄畅园的最大特色，而这一借景效果正是通过巧妙利用园址形势、形成隔水观山的格局而最有效地引导获得。

在东岸西望获得真山绵延景致效果的同时，这种隔水观山还呈现着强烈的画意效果。前述“张氏之山”的“平冈小阪，陵阜陂陁”，本身就以富于画意而著称，尤其是元代画家的画风。在寄畅园的这个实例中，张鉽除了对假山本身形态的改造，还以山与水的配合，以及建筑作为赏景场所的设置来获得画意效果。从知鱼槛西望，隔水山景，正似元代画家，尤其是倪瓒所常用的“一河两岸”式构图风格。虽然实际距离不远，却以滩、矶、湾、桥、树、坡等形成多样层次。清代文人杨抡《芙蓉湖棹歌》中“秦园妙处知鱼槛，一笔倪迂画里秋”之句，[35] 以及乾隆年间状元石韫玉《游梁溪秦氏寄畅园》诗中“留得云林画意多”的称颂，[36] 正是对这种倪瓒画风效果的欣赏。

与此同时，水畔山景也因狭长池面而沿南北长向延展，如果左右转目而视，或在池东廊中行进游观，对岸山水则如一幅手卷横呈、徐徐展开，这或引人作《富春山居图》长卷般的遐思。邵长蘅《惠山游秦对岩太史寄畅园六首》中有“陂陀子久画”的比喻，[37] 正是将隔池山景与这种黄公望（字子久）的画风相联系。这里也可以看到，“张氏之山”的画意并不拘泥于某一位画家的风格，并非某种画面形式的直接模拟，而是更在于能引发多种联想，在妙不可言的画意效果中获得杰出山水画所追求的“象外之境”意蕴。

其次，除了东西横向的隔池观赏，南北纵向上也可以获得平远丰富的层次效果（图 11-图 12）。水池中段伸出“鹤步滩”及其上向水面伸展之树，[38] 与东岸伸向水中的“知鱼槛”相呼应，使池面在此收缩，形成有分有聚、有收有放的层次变化，景象呈现得格外生动有致；而在北侧，又有“宛转桥”斜向跨水而划分池面，并使池面在此稍作弯折。[39] 从而，在水池南北两岸处观景，亦可有某种层次丰富的画意；而且由于池面收束与划分并非均一，大致南侧以聚为主，北侧则着重于散，从南北两侧所对望而得的远近效果又有差异，景色的层次和深度更显丰富。在欣赏纵远水景的同时，一侧山景与之又能密切结合，形成山水共生的画面。而在北岸屋前平台，沿水面南望还可以远眺锡山及其上的龙光塔，如华硕宣《暮春同曾梅峰游秦园》中"青川塔影浮"的效果，[40] 是另一极佳借景，甚至成为今日寄畅园的标志性景观。

可以看到，清初张鉽改筑中，对于山池主景删繁就简，去除了以往晚明时作为重点主景的大量建筑亭廊，仅以池东少量亭廊与池西假山相对，再辅以桥梁、滩矶等些许变化，而使山池主景从繁复而离散变为简练而整体，以横、纵两个方向的配合将"山高水远"的效果展现得淋漓尽致，[41] 又以巧妙的园外借景来大大拓展了园中景象，更以山水相依、人景相对而获得丰富而强烈的画意。可谓用笔极简而成效极佳，大气磅礴而意蕴丰富，而与晚明时的效果迥异。

4 "如入岩谷"：谷涧的游观体验

寄畅园历来以引入二泉之水、形成溪涧泉瀑之景为傲，作为园中一大特色。园景初创时期，秦金《筑风谷行窝成》诗中有"曲涧盘幽石"之句，[42] 可见曲涧之景。秦瀚、秦梁父子时期进行了改筑，秦梁《山居赠答》中亦有"白知泉瀑泻"的泉瀑之景叙述。[43] 至秦燿时期，由各种文字、图像，我们对此景的了解更为详细，尤其是王穉登《寄畅园记》的描述："台下泉由石隙泻沼中，声淙淙中琴瑟……引'悬淙'之流，甃为曲涧，茂林在上，清泉在下，奇峰秀石，含雾出云，于焉修禊，于焉浮杯，使兰亭不能独胜。曲涧水奔赴'锦汇'，曰'飞泉'，若峡春流，盘涡飞沫，而后汪然渟然矣。"[44] 对此，我们又有宋懋晋的绘图，"曲涧"（图 2）及"飞泉"（图 5）是重要园景所在，后者也成为涵碧亭的主要对景（图 4）。可以看到，园外的泉水引入之后先汇于小沼，然后在假山之上设宛转曲涧，最后形成飞瀑而入锦汇漪。对于涧瀑的欣赏，除可观景，也可听音，如秦燿《园居》诗中“静听水淙淙”，[45] 王穉登《寄畅园》中的“泉声虢虢石磷磷”，[46] 邹迪光《寄畅园五首》中的“流泉代鼓弦”[47] 等等。如王穉登《寄畅园记》所言，当时寄畅园的特色“其最在泉”，这股涧泉成为此园特色的最重要体现。

在清初张鉽的改筑中，这一特色景观也得到了大刀阔斧的改造，在保留涧泉、飞瀑[48] 特点的同时，更增加了新的内容，将山上的“溪涧”，改造为山中的“谷涧”。将秦燿时的复原平面图（图 1）与当代寄畅园平面图（图 9）进行对比，可以看到涧泉入池的位置发生了明显改变。这表明，这一涧泉得到了重大改造，有新的涧道得到开辟，而非仅是旧涧的整修。这一结论也可以得到文献的支持。与张鉽同时代的许缵曾《宝纶堂稿》中提到这次改筑：“吾郡张鉽以叠石成山为业，字宾式。数年前为余言，曾为秦太史松龄叠石凿涧于惠山。”[49] 这里张鉽自述的对寄畅园的改筑活动，“凿涧”正是这里的涧泉营造，一个“凿”字形象地说明了这一工程的要点：在以土为主的假山之中新开凿出一条新的谷涧。类似的还有黄与坚《锡山秦氏寄畅园》一文中的“引慧山之水汇为池，破为涧”，[50] 也说明这一改筑有着将假山新“破”开而“为涧”的重要工程。在将山体“破”出涧道、“凿”出空间的同时，还有在谷道两侧叠石的工程，前述余怀《寄畅园闻歌记》中有“倚峭壁而听响泉”，乾隆时张英《南巡扈从纪略》有“叠假山为溪谷”，[51] 正是说明这一溪谷的形成还伴随着峭壁假山的营造。从今日遗存的“八音涧”（图 13），可以看到这一工程的规模和效果，随曲涧延伸的一人多高的岩壁营造，在整座以土山为主的假山中有着规模巨大的用石量，从工作量角度看甚至很可能成为整个改筑工程的重心。

另一则新近发现的史料进一步表明了这一改造的意义。钱肃润《夏日雨后秦子以新招过寄畅园观涧水》诗后的一段说明中有：“南垣治奉常东园，创用黄石，朴素自然，为累石开天之祖。然娄东无山，不能致飞瀑潺湲，耳根为一大恨。

今小阮补之寄畅，遂称完美。”[52] 其中将寄畅园改筑与张南垣早年对太仓王时敏（“奉常”）的乐郊园（即这里“东园”）进行了对比。当年的乐郊园改筑是张南垣年轻时的成名作，[53] 这里“创用黄石，朴素自然，为累石开天之祖”的赞誉可见地位之高，然而因为太仓的平坦地势无法营造出涧泉飞瀑之景，成为遗憾；现在寄畅园的优越地形条件，完全可以将新的造园叠山理念用于其中。由于当时张南垣已经年迈，于是派其侄儿（“小阮”[54]）张鉽前往，终于营造出理想的山水涧瀑效果，弥补了张南垣长久的遗憾，而“完美”实现此生的夙愿。

那么通过这番改造，与寄畅园此前已有的涧瀑之景相比，这一新凿的涧道要优越在何处？

康熙《无锡县志》中秦松龄本人对此景的描述为：“引二泉之流，曲注层分，声若风雨，坐卧移日，忽忽在万山之中。”[55] 这里词句不多，却可以细细分析。“引二泉之流”为以往一直有的特色，这里得到沿用，是水景的基本来源。前述此处涧泉水景的欣赏有两方面：一是视觉之态，一是听觉之音，二者在改造后都得到了加强。

“曲注层分”，是景观形态，其中又可分为两方面。一是“曲注”，可视为平面上的弯曲形态设计，以前就以“曲涧”为景名，这里对此承袭，而从李念慈《奉陪杨退菴中丞夜集梁溪秦留仙太史园林酣畅达晓，明日重过，遍观池涧亭榭诸胜即事赋赠》诗中“曲折迴岩坳”之句，[56] 则展示出在两侧石壁高出的“岩坳”空间中的一种幽深感，呈现出与普通涧流所不同的新意；而在“曲”的同时，从姜宸英《惠山秦园记》中“中叠石条为细涧分流”[57] 以及清高宗《寄畅园再叠旧韵》中“瀑出峡分数道潺”，[58] 还可以看到其中又有分为数道细涧的变化，这是平面形态的新发展。二是“层分”，则是竖向上的层次形态设计，这一改造后的泉涧又被称为“三叠泉”，[59] 见于吴绮《寄畅园杂咏（九首）》[60] 和秦道然《寄畅园十咏》等，[61] 可以看到，与以往“曲涧”之名相比，“三叠泉”更突出竖向变化特点，可见通过“层分”营造，形成了新的标志性特色。

“声若风雨”，与以往的“虢虢”“淙淙”形容相比较，说明在声景方面更加突出——由于谷涧空间对水声的汇聚，效果得到加强；这也并非仅是主人的自夸，钱肃润《夏日雨后秦子以新招过寄畅园观涧水》诗中“飞泉喷激响潺潺，万顷波涛一涧间”，[62] 这里更有“万顷波涛”般轰鸣的夸张形容。而从邵长蘅《游慧山秦园记》中的“泉虢虢石罅中，鸣声乍咽乍舒，咽者幽然，舒者淙然，坠于池潀然渊然”，[63] 还可以看到，除了声景效果的放大，在涧流的各段还通过多样化的处理而使泉声产生细腻多样的效果。

而这一改造最重要、最精彩的效果，则在“坐卧移日，忽忽在万山之中”。如前所述，“张氏之山”重视身体进入的真山水游观体验。晚明寄畅园的涧瀑之景尽管出色，却仍只是外在之“景”——虽然溪涧可以到达，但主要还是通过悬淙亭赏涧、涵碧亭观瀑，而缺乏人在景中的游观体验之“境”的获得。而通过新凿高谷深涧，将游人置身其内而游赏，可以到达前所未有的身临其境的峡谷山水之感。这里的“坐卧移日”表达了游人在内的动态游观，“忽忽在万山之中”则正是所要追求的真山水境界；秦松龄在《无锡县志》中所归纳的张氏造园“宛然天开”的效果，也得到极好的例证诠释。

从而我们可以理解，为何以往的涧瀑已经相当动人，却仍要花费大气力进行新的大规模改造；经过张鉽改筑的谷涧，确实可谓脱胎换骨，不仅在形态和声景方面的效果更加显著，更是前所未有地达到“忽忽在万山之中”的真山水内游观体验，而这正是时代造园新意的一大标志性追求，在寄畅园得到了完美实现，这一谷涧之境也与主景山池画意一起成为寄畅园的两大标志性园景。

今日的寄畅园“八音涧”，与清初改筑后的效果有所差异，入池飞瀑已基本消失，水平向的“细涧分流”和作为“三叠泉”的竖向“三叠”“层分”也已几乎不见，声景上更与“万顷波涛一涧间”的汹涌与轰鸣相距甚远。这些既是由于所引入二泉之水流量的不足，更与此后历代的维修有关。陈从周在《续说园》中提到“无锡寄畅园八音涧失调，顿逊前观，可不慎乎？可不慎乎？”[64]也对近世的不当修缮感到遗憾。但总体而言，还是保留了相当多的当年效果，这道长三十余米的黄石涧峡，空间上大体保持了曲折高下、阔狭旷奥的变化，仍有不绝于耳的汩汩潺潺之声，游观中还是颇有深山幽谷之趣。

5 结语

作为十七世纪江南造园新风引领者的张南垣及其传人的高超山水营造技巧成就了寄畅园这一杰作，而寄畅园的优越条件也为张氏造园成就了一件可与张南垣早期成名作乐郊园并称的代表作品。结合文献中“张氏之山”的特色以及与其他时段（尤其晚明与当代）园貌的关联比较，可以较为清晰地认识清初寄畅园山水改筑的两大杰出成就——视觉效果上的真山之隅与山水画意，以及游观体验上如入岩谷的真山之境；虽历经沧桑变迁，在今日寄畅园中仍能真切感受。

在寄畅园成功得到时代造园新法改造的同时，也为我们进一步理解“张氏之山”及十七世纪江南造园变革提供了极佳范例。以张南垣为代表的十七世纪造园新法中，不仅是“平冈小阪，陵阜陂陁”的富有新意的山水形态，更在于深层的画意追求；而这画意又不仅仅在于视觉上对山水的美妙品赏，还在于“拟入画中行”的真山水游观体验——通过寄畅园改筑，这在山池的画意和谷涧的游观中得以神奇实现；因而时人会将此园与作为张南垣开创时代造园新意的乐郊园相提并论，以“遂称完美”来表达赞美之情。寄畅园虽然是江南仅存的“张氏之山”所在，却也可以成为“张氏之山”的代表作；从中所呈现的十七世纪中国园林山水营造的高超艺术水准，至今仍是未被逾越的高峰。

"Knowing the Painting Vein" and "Being Like in a Ravine": Jichangyuan Landscape Reconstruction in Early Qing and "Zhang's Mountain" in 17th Century Jiangnan[1]

GU Kai

1 Introduction

Jichangyuan in Wuxi has always been a well-known garden in the Jiangnan area. It received praise when it was first completed in 1527, then more compliments during the later Ming period (1368–1644), and became widely admired during the Qing period (1616–1911). The garden acquired even greater fame after repeated visits by the Emperor Kangxi (1654–1722) and the Emperor Qianlong (1711–1799), who delighted in it so much that he not only wrote many poems about it, but also emulated it in his Qingyiyuan (today's Summer Palace) in Beijing. Over time, Jichangyuan underwent a number of modifications. But despite this, it has remained, to this day, a magnificent example of the highest form of Chinese garden art, and in many experts' eyes it is the finest of its kind.[2] The current form and appearance of Jichangyuan is largely attributed to the last major reconstruction, which took place during the early Qing period, ca. 1667.[3] This early Qing reconstruction is "key" to understanding the significance of the garden landscape, and therefore, deserves careful study.

An abundance of texts and images exists on the previous reconstruction during the late Ming of the 1590s,[4] but primary materials on the early Qing period reconstruction are very limited. For example, a record does exist within the local chronicle of *Wuxi xian zhi*, compiled in the Kangxi period (1662–1722). In this text—believed to be written by the garden owner of the day, Qin Songling—one sentence reads, "When the garden (reconstruction) is accomplished, all those previously regarded as famous spots are discarded."[5] This first-hand account tells us that the original appearance of the garden was greatly transformed during the early Qing reconstruction. However, it does not help us to understand how this reconstruction was carried out. So while contemporary understanding of the transformation of Jichangyuan during the late Ming is rather rich and clear, that of the early Qing is still relatively weak and vague. Among contemporary studies on the reconstruction of Jichangyuan in the early Qing, Huang Xiao has made the major contribution to confirmation of dates, identification and interpretation of the main landscape and garden features and ideas, as well as offering speculations with regard to the methods of construction.[6] This work constitutes the foundation for further study. Qin Zhihao has also contributed research on the historical context of the early Qing reconstruction, this being the latest progress.[7] Nevertheless, present studies have not yet provided a satisfactory answer to how the early Qing reconstruction caused "all those previously regarded as famous spots" to be "discarded."

To remedy this shortage of documentation, we may turn to historical records and related studies. Again, in the chronicle of *Wuxi xian zhi* of the Kangxi period we read,

> Mr. Zhang Lian, also known as Nanyuan, from Yunjian, can make garden rockery as layered mountains and deep valleys, just like natural creations. He has changed the previous stereotype completely and formed his unique style. His technique has been valued greatly in the world for decades, but it has not appeared in Wuxi before. For this project, Zhang Lian sent his nephew Zhang Shi to try his best to carry out the work.[8]

Thus we learn that the project was undertaken by Zhang Shi—an important heir to the craft of Zhang Nanyuan, who's work he carried on. His uncle Zhang Nanyuan was a garden master and artist of great renown. He was admired for his skill, innovative techniques, and painterly style. Zhang Nanyuan's great influence and the direct connection between the master and Zhang Shi prompts us to consider the early Qing reconstruction within the wider context of the history of Chinese garden art, and especially the transformation of this art in 17th century Jiangnan.

Present scholarship abounds on Zhang Nanyuan (1587–1666)[9] as well as his garden rockery style, known as "Zhang's mountain,"[10] but his built works have all vanished in history. The landscape of Jichangyuan, as reconstructed by his nephew Zhang Shi, is the only heritage of "Zhang's Mountain" that remains preserved in Jiangnan. By studying written records of "Zhang's Mountain" we can better understand the early Qing reconstruction and the significant features of Jichangyuan landscape. Furthermore, together with historical materials and the present condition of the garden, a study of Jichangyuan can contribute to a new understanding of Zhang's style of garden making, as well as more broadly, the history of the Chinese garden in 17th century Jiangnan.

According to available records, the early Qing reconstruction was the only major work carried out on the main garden landscape since Qin Yao's earlier reconstruction of the 1590s. Therefore, it is possible to study the early Qing reconstruction by comparing the features of the garden during the late Ming with those that we find in later texts and in the appearance of the garden today. Although considerable alterations and even damage occurred since the early Qing reconstruction alterations were made to buildings and planting—the core structure of the garden resisted transformation. This main organising principle of the garden is the landscape feature called "mountain-and-water" (*shan-shui*). This feature can be still observed today as it was established in the early Qing, and its effects can be generally understood from the contemporary plan and pictures of the garden.

2 "Zhang's Mountain": the "Avant–Garde" of Garden Making Art in 17th century Jiangnan

During the 17th century, in late Ming Jiangnan, an important transformation was taking place within the Chinese tradition of garden making. Zhang Nanyuan was the leader of the new direction, and widely acknowledged to have been the most important exponent of the art.[11] The most detailed research on his work and reputation has been carried out by the historian Cao

Xun, whose account gives an indication of the significance of his influence:

> Zhang Nanyuan is the principal master of garden and rockery making in China. Throughout the entire history of this tradition, his work is unsurpassed in quality and quantity. With his invention of a new style and his great contribution he opened a new era within the history of Chinese garden and rockery making. His achievement pushed Chinese garden and rockery art to its peak, and he was an immense influence to both his contemporaries and to later generations.[12]

Zhang Nanyuan acquired fame for his work in the late Ming of the early 17th century and continued to gain wide acclaim through numerous projects during the early Qing of the late 17th century. His sons, having followed him into the garden making business, inherited his skill and reputation and carried on his legacy.[13] His greatest innovation was his style of garden rockery, dubbed "Zhang's Mountain," in Lu Yanzhe's *Biography of Zhang Taoan* for Zhang Ran (Zhang Nanyuan's youngest son): "Mr. Zhang Nanyuan excels in artificial rockery making. His work has spread throughout the country and among noble people. When people see his work, they know it is 'Zhang's Mountain' without asking."[14] A statement like this attests to Zhang's achievement and the hallmark of his style.

To gain a clearer understanding of the distinctive quality of "Zhang's Mountain" we may consider its visual form and the bodily experience it offers. In his analysis of the visual form, Cao Xun identified four main characteristics. The first is "the application of the idea of landscape painting within the practice of making gardens and rockery." The second is "an opposition to the unusual rocks and mystifying caves that had been in vogue before. Instead, he proposes gentle slopes, segmental brooks and valleys, scattered flowers and rocks in a randomly arranged way." The third is "proposing artificial mountains by combining earth with rock." And the fourth is "considering the overall garden composition, with mountains, water, buildings, and planting all arranged organically."[15] The first characteristic describes the "painting-idea" (*hua–yi*)—a term frequently used in late Ming texts on garden making which constitutes the basic principle of the style. The latter three are the concrete realization of this principle.

The "painting-idea" was the deep driving force of the new direction and Zhang Nanyuan became its leader.[16] Evidence of this is found in Wu Weiye's (1609–1672) *Biography of Zhang Nanyuan*, where he writes that, "Mr. Dong Qichang and Mr. Chen Jiru complimented him greatly for his work: the mountains of Jiangnan bear rocks among earth, as Huang Yifeng [Gongwang] and Wu Zhonggui [Zhen] often said; this shows his work knows the 'painting vein' (*hua–mai*)."[17] Being the cultural leaders of the time, Dong Qichang (1555–1636) and Chen Jiru (1558–1639) recognised the painterly philosophy of the Yuan era painters Huang Gongwang and Wu Zhen in the work of Zhang Nanyuan. Wai-Yee Li explains that the inner connection between Zhang's garden making and Yuan paintings lies in that "Zhang Nanyuan inherits the paintings' aesthetics of reticence. Rejecting the use of huge rocks of curious shapes at great expense, he instead favours the sense of movement—'the veins of rocks'—strategically placed rocks on low mounds."[18] Cao Xun also further summarised the "painting idea" with more relevant historical texts, "the painters Huang Gongwang, Wang Meng, Ni Zan, and Wu Zhen provided the inspiration for Zhang's manner of making artificial mountains and rivers, winding and meandering into remote and distant atmospheres that surpass nature."[19] This affinity with the painting of the Yuan period departed from the previously popular style of making large and grand rockery, full of extraordinary shapes and maze-like caves. Instead, it sought the concise methods, natural effects, and imagination "beyond representation" that Yuan painting invoked.[20]

Aside from the influence of Yuan landscape painting on the visual form, another important aspect of how "Zhang's Mountain" departs from the previous style is the greater emphasis of combining bodily experience with the visual effect of the "painting-idea."[21] In Dai Mingshi's (1653–1713) *Biography of Zhang's Family* he writes that Zhang Nanyuan's "garden making has ingenuity ... Although inside a tumultuous city, people can feel as though they are in a ravine."[22] In a book by Lan Ying and Xie Bin we find the note "for a land of half mu, after his work, people get the feeling of residing in a deep valley."[23] The words "being like in a ravine" and "feeling of residing in a deep valley" give emphasis to the bodily and physical experience of being in the garden landscape of "Zhang's Mountain."

There is an inherent relationship between the visual form and the bodily experience, and both constitute the manifestation of the "painting-idea." I have argued elsewhere that the idea of landscape painting is not just about formal appreciation. When the "painting-idea" was established as the central principle in garden making, both the creation and appreciation of the garden were deeply affected. In the treatise of *Yuanye* (*Craft of Gardens)* by Ji Cheng (b.1582) the expression "wandering within a painting" is vivid and influential.[24] This "wandering of the mind" became important and immanent within the tradition, and led to the foregrounding of the experience of actually moving through the garden.[25] In his version of *Biography of Zhang Nanyuan*, Huang Zongxi (1610–1695) wrote, "the painting effects of naturalness of Jing Hao, simple elegance of Guan Tong, variety of Mi Fu, desolation of Ni Zan—all can be physically immersive and experienced through the body."[26] This is a pertinent expression of the strong relationship of "knowing the painting vein" and "being like in a ravine" in Zhang Nanyuan's work.

With this understanding of the basic features of "Zhang's Mountain," we can study more effectively how the early Qing reconstruction of Jichangyuan proceeded and how "all those previously famous spots" were transformed. In other words, what was changed and why the change was for the better. As the title of "Zhang's Mountain" implies the main feature of Zhang's work that Jichangyuan is famous for is its "mountain-and-water" scenes. Therefore, our focus here is mainly on the reconstruction of the garden's "mountain-and-water" elements.

In poems written bout Jichangyuan after Zhang Shi's reconstruction, two landscape areas are often mentioned with special attention. These focal points are evoked in verse as "the mountain greenness approaches the curtain; the water flows with swirling and trickling in rock crevices" by Shao Changheng,[27] and "watching the still water while walking along the long gallery; listening to the resounding spring while leaning on the cliffy stones" by Yu Huai.[28] From these texts we know that one is around the main pond area—with "still water" in the pond and the mountain scene of "greenness" to one side. This scene is to be appreciated from the galleries at the other side of the pond. The other is a gully with a stream on the side of the garden mountain, with "swirling flows" and "resounding spring," being appreciated together with "rock crevices" and "cliffy stones." Therefore, a study of the landscape reconstruction can be carried out on the basis of these two areas: the combination of"mountain-and-pond," and the mountain stream. These two parts correspond to the two features of "Zhang's Mountain."

3 "Knowing the Painting Vein": the Reconstruction of "Painting-Idea" of "Mountain-and-Water"

The main garden pond, named "Ripples of Brocade Converging," is the main scenic area of the whole garden. At its western shore there is an artificial mountain, while along its east bank there is a group of pavilions and walking galleries. The composition of "mountain-and-water" forms the basic scenic structure of the whole garden.

Very often in traditional garden making, the construction of the main scenic feature of "mountain-and-water" is the most important part of the project. In many famous Jiangnan gardens of the late Ming, such work cost vast amounts of money, labor, and time. Many projects took more than a decade to complete.[29] However, immediately after the damage of the war,[30] the economic condition in the early Qing period was rather poor. The reconstruction of Jichangyuan during this period took only a little over one year for the whole project.[31] We may ask, how could the reconstruction transform the appearance of the garden so much, and achieve the wonderful effect within such a short time frame and at a low cost?

The success of the project lies in its brilliant design by the master craftsman. In comparing the garden before and after reconstruction we find that there was little overall change in the position, size, and arrangement of the mountain and the pond themselves. However, what did change significantly was the layout—especially the relationship between landscape and buildings.

Extensive studies by Huang Xiao and other scholars have been conducted on the appearance of the garden after Qin Yao's construction in the late Ming. These studies have led to a plan (Figure 1) drawn according to rich historical evidence. We know that the main "mountain-and-water" area was arranged in a complex composition with abundant "scenic spots" (*jing-dian*). Within the mountain area facing the pond, there were "Winding Stream," "Pavilion of Gurgling Water," "Terrace of Refreshing," as well as many elaborate rocks (Figure 2). The pond had a very close relationship with the buildings. A long walking gallery crossed the water from west to east, and mid-way along it a pavilion named "Rail of Knowing Fish" was sited the middle of the pond (Figure 3). This pavilion and the walking gallery divided the pond in two parts—north and south. In the south, the "Rail of Knowing Fish" embraced the water, with the "Pavilion of New Moon" and "Studio of Rising Cloud" connecting the long gallery to the west and"Rose Screen"to the east. In the north, the "Pavilion of Containing Greenness" was positioned on a small isle (Figure 4) that formed the core and acted as a place for watching both the mountain and waterfall in the west (Figure 5). It also acted as an opera stage if observed from the "Rail of Knowing Fish" or the "Studio of Pure Bamboo Fence" in the west.

A recent study also shows that "Jichangyuan in the late Ming was indeed a luxurious garden" with multiple exquisitely decorated buildings.[32] With such plentiful and magnificent buildings arranged within the landscape, visitors could appreciate a variety of landscape scenes comfortably—both in moving and resting. Such a landscape arrangement—with complex architecture, ornate and elaborate rocks and caves in the artificial mountain, as well as the very fine buildings—was in line with garden aesthetic that people admired—the widespread garden fashion of the time.

However, in the great change of garden aesthetics in the 17th century, such previously admired constructions became problematic. The new attitude concerned not only the artificial mountain, but also the effect caused by the elaborate buildings. Suddenly, the long walking gallery, as well as the connecting buildings, not only divided the pond and made the water scene seem less attractive, but also drew a separation between the artificial mountain and the pond, weakening the important relationship of "mountain-and-water." The feeling of the natural landscape of "mountain-and-water" was felt to exist only sporadically (for example at the waterfall). The new aesthetic movement pursued simplicity and elegance of natural landscape in affinity with the idea of landscape painting in the 17th century. It considered the old layout inappropriate because the overall natural effect was seen to be fragmented by excessive buildings and heavy decoration.

Now let us compare the above with images of the garden in the 18th century (Figures 6–8) and the garden plan surveyed in the 1980s (Figure 9), which mainly correspond to one-another. The artificial mountain no longer has elaborate rocks or caves. Instead, it is mainly made of earth with more natural-looking rocks dotted sparsely with lush plants—more similar to a real mountain foot. This corresponds to Gu Jingxing's annotation to Shao Changheng's poems, "the artificial mountain has no rocky peaks but the form of extending slopes."[33] The pond is now vast open space, with no architectural construction on the water or between the mountain and the pond. A group of pavilions and a gallery to the east of the pond allow the mountain to be appreciated from across the water. At the north part of the pond there is a bridge that connects two banks.

Comparing the appearance of "mountain-and-water" before and after the 17th century, we can deduce the main actions of the reconstruction. The artificial mountain was transformed according to the typical "Zhang's Mountain" method. Mainly based on the previous landform, "gentle slopes" of earth mound dotted with rocks, trees, paths, and a pavilion at the back were made in accordance with the landscape "painting-idea." The pond area saw the most drastic transformation. All the buildings were removed from the pond and the area where the mountain and water connected. Even the islet and pavilion on it were abandoned. A new more naturalistic effect was established at the west shore of the pond, made as though with the mountain foot extending into the water. On the opposite bank the only building group of pavilion and gallery faced the "mountain-and-water" connection. In this way, the pond scene became more prominent with a vast water surface. Thus, the relationship of "mountain-and-water" was greatly enhanced and brought into a greater intimacy of relation.

This new design contrasted with the previous appearance of the garden. The strong effect of the new construction can be analyzed following the length and the breadth of the pond.

From the breadth of the pond, in the east-west direction, there is a clear composition of the sequence "mountain-water-building." The complex of pavilion and gallery at the east bank of the pond such as "Rail of Knowing Fish"[34] is an important place for appreciating the water scene and watching the mountain across the water (Figure 10). Such an arrangement also takes full advantage of the garden located at the east foot of the Hui Mountain. The artificial mountain in the garden is seen as though it is a continuation of the real mountain outside the garden. This makes for an ideal use of the "borrowing view" (*jiejing*) idea: observed from the east pavilion and gallery, the artificial mountain appears as though stretching into the distance, merging with Hui Mountain in the back.

From this perspective, the garden boundary vanishes and the garden landscape is greatly extended to the broad universe outside, which is precisely the ideal realm pursued by the literati garden. In the long poem of *Qin's Garden* by Yuan Mei (1716–1797) the opening lines read, "To make a tall construction, it must be based on a foundation. To make a garden, it must be based on something too. How beautiful is Qin's garden! It borrows the Hui Mountain!"[35] In Yuan Mei's eye, the most important feature of Jichangyuan lies in this "borrowing-view," constituted by the utilization of the topographical condition and the layout that enables one to observe the mountain from the buildings across the water.

As previously mentioned, the making of "Zhang's Mountain" is well-known for the manner of employing visual features connected with the "painting-idea" (especially of Yuan painters). In many poems written about the main mountain-and-water scene in early-Qing Jichangyuan, it is this painting effect that is often praised. For example, "the wonderful scenes of Qin's garden can be obtained at 'Rail of Knowing Fish,' just like the strokes in a Ni Zan's painting on the autumn" by Yang Lun in the 18th century,[36] and "much of Ni Zan's 'painting-idea' is acquired" in Visiting *Qin's Garden* by Shi Yunyu (1756–1837).[37] Today, we can still feel that the powerful painterly effect of the view of the mountain across the pond is similar to the conspicuous style of "one river, two banks" in Ni Zan's paintings.[38] While the real distance is modest, multiple "layers" formed by various arrangements of shoals, bays, slopes, rocks, trees, and bridges seem to extend the space into the distance.

At the same time, the same "mountain-and-water" scene unfolds along the north-south axis of the water's edge. If one were to observe it from the across the pond while moving—shifting the gaze or walking along the gallery—it would reveal itself like a handscroll opened gradually. This experience may remind one of the long scroll of *Dwelling in the Fuchun Mountains* by Huang Gongwang. In one of Shao Changheng's poems, this scene is related to Huang Gongwang's painting style in the line "slanting slopes just like Zijiu's painting."[39] These comparisons to a variety of painting ideas employed in the early Qing reconstruction demonstrates that the painting idea of "Zhang's Mountain" is not limited to a specific painter's style, nor does it directly imitate a specific painting form. Rather, it aims at arousing the imagination and pursues the "realm beyond image" (*xiang wai zhi jing*) as the highest expression in art, both in painting and garden making.

Along the length of the pond in the north-south direction, a landscape effect of layering and remoteness takes hold (Figures 11–12). In the middle part of the west bank the "Shoal of Crane Walking" together with a slanting tree of luxuriant foliage[40] extends into the pond, echoing the new pavilion of "Rail of Knowing Fish" and protruding into the water. This makes the surface of the pond appear to shrink at this point, and helps to create a more varied appearance of water scenes and layers. In the northern part of the pond, a bridge divides and slightly turns the water surface.[41] In this way, painterly effects with rich layering are achieved on both banks, in the south and the north. As the variation of tightening and division of the water surface is not even— the south part is larger and has the effect of gathering, while the north part is slightly smaller by comparison and appears to disperse around the perimeter—the effects of layering and depth are differentiated between the views from the south and the north respectively. This makes the garden scenes appear more diverse and attractive. Another layered view of the water can be appreciated together with the mountain scene, but this time from a different perspective. Observed from the terrace in front of the building on the north bank of the pond, a distant view of Xi Mountain with a pagoda can be achieved above the water. This is described as "pagoda reflection floats on green water" in Hua Shuoxuan's poem.[42] This is another fabulous "borrowing-view" in the garden that has become iconic in Jichangyuan.

In summary, the new aesthetic approach of early Qing made it possible to employ a range of sophisticated techniques in a much more economical way than the previous way of late-Ming. Economical—because the specific features that had been multiplied to great expense before were now attenuated and allowed to be more naturalistic, rather than striking and exotic. This made sourcing and construction cheaper, as well as more reserved with fewer extravagant buildings. This gesture of reduction and removal in fact permitted the opening up of the landscape itself to the effects of layering, borrowing of views, unfolding through movement like a painting scroll, as well as of bringing the pictorial and painterly effects together with the bodily experience through movement and touch. The experience of garden in early Qing thus became enriched, enlarged and enlivened. The new aesthetic direction supported by Zhang's brilliant use of techniques in garden making, brought the experience of nature as the realm of "imagination beyond image" to life.

4 "Being Like in a Ravine": the Touring Experience Inside the Valley in the Mountain

One constant feature of Jichangyuan is the scene of flowing water that courses from nearby "Second Spring under Heaven" entitled by Lu Yu (ca. 733–804), and worshiped as "Tea Saint" in history. Soon after the garden was first created in the 1520s, a poem by Qin Jin depicted the scene of the stream with the words "the winding stream twines about quiet rocks."[43] In the 1560s after the first reconstruction, a waterfall scene appears in Qin Liang's poem as "whiteness is pouring from spring and waterfall."[44] By the 1590s, we come to know the scenes much better through to multiple texts and images. In *Record of Jichangyuan* by Wang Zhideng (1535–1612),

> ... the water from "Pavilion of Hanging Currents" is constructed as a winding stream. With lush woods above and clear spring under, fancy rocks with mist around, the activity of "waterside ritual with floating wine cups" can be held [...]. The stream water, titled "Flying Spring,"rushes to the pool of"Brocade Converging."The water is like through a gorge in a spring with swirling flows and flying sprays, then it comes to a vast expanse and becomes peaceful.[45]

In addition to this text, paintings by Song Maojin, "Winding Stream" (Figure 2) and "Flying Spring" (Figure 5) also present us with the scenes of spring and waterfall, with the waterfall scene of "Flying Spring" also being the main view in "Pavilion of Containing Greenness" (Figure 4). According to these texts and images, the spring water brought into the garden from the outside is first channeled into a small pool in a high place. Then, a winding stream is laid out in the artificial mountain, and finally, through a waterfall, water flows into the main pond. But the experience of "stream and waterfall" is not only about the view. It is also an aural experience. There are many verses on the soundscape of the water, such as "Listening quietly, water is gurgling" in Qin Yao's poem of *Garden Dwelling*,[46] "the spring sounds gurgling while rocks look shining" in Wang Zhideng's poem *Jichangyuan*,[47] and "flowing spring sounds from zither strings" in *Five Poems on Jichangyuan* by Zou Diguang (1550–1626).[48] With such an attractive landscape effect, the stream scene plays an important role in the appreciation and experience of
the garden landscape. Wang Zhideng even wrote in his *Record of Jichangyuan* that "the most conspicuous feature lies in the spring" of this garden.

In Zhang Shi's early Qing reconstruction, this characteristic landscape was drastically transformed. The stream on the surface of the artificial mountain became a much deeper gully. Comparing the hypothesised plan of the garden of Qin Yao's time (Figure 1) and the plan of the garden surveyed in the 1980s (Figure 6), we can clearly see that the location where the water flows into the pond has changed. This is evidence that the stream was significantly transformed, and a new gully was constructed rather than being rebuilt in the previous site. This conclusion is also supported by the historic literature. Xu Zuanzeng, a contemporary of Zhang Shi, narrated in his *Drafts of Baolun Hall*, "Zhang Shi, courtesy name as Binshi, also from my county, takes rockery making as his profession. He told me several years ago that he made rockery and chiseled a valley for Mr. Qin Songling at Hui Mountain."[49] Here, Zhang Shi's own words describe the reconstruction of Jichangyuan. The specific word "chisel" is a vivid allusion to the key of the project: to open a new valley in the artificial mountain, which is mainly made of earth. Another piece of text of *Jichangyuan of Qin Family of Xishan* (1668) by Huang Yujian reads, "water from Hui Mountain was guided into the pond, a ravine was made by breaking (the garden mountain)."[50] This tells us that the task of "breaking" the artificial mountain into a ravine was an important part of the project. At the same time as "breaking" into a ravine and "chiseling" out space, the rockery making on both sides of the valley was also a big project. The gully was formed by the making of cliff rockery, which is known from Yu Huai's line, "listening to the resounding spring while leaning on the cliffy stones" as mentioned above, as well as "rockery made to form a stream valley" in *Brief Record of Accompanying Emperor's South Touring* by Zhang Ying in the period of Qianlong.[51] Finally, we can begin to understand the scale and effect of this project from direct experience of the remaining "Gully of Eight Tone" (Figure 13—its craggy cliffs rising above the height of a person and extending along the long winding gully. Considering the immense quantity of rock utilised here, this part of the project may be the key part of the whole reconstruction from the perspective of workload.

Another piece of the text gives further evidence of the significance of this transformation. In one poetic story told by Qian Surun, he wrote, "when Zhang Nanyuan made the garden of Wang Shimin, he made creative use of the yellow stone [to make the rockery] plain and natural, and can be considered the pioneer of such rockery making. However, there are no natural mountains in Taicang and a waterfall scene cannot be made, which made him regretful. Now he sent his nephew to realize it in Jichangyuan, so the perfection can be achieved."[52] Here the reconstruction in Jichangyuan was compared with Zhang Nanyuan's early famous work of Lejiao Garden in Taichang.[53] Due to the finer and more suitable topographic condition of Jichangyuan, here Zhang Nanyuan's new rockery-making idea could be realized with the compelling scenes of "ravine and waterfall," making up for the regret that he could not achieve a perfect effect in Lejiao Garden on flat terrain.

Now we may ask: how is this new gully superior to the previous scene of stream and waterfall?

To to answer this question we may analyse the words of the garden owner Qin Songling in detail. He writes, "The water of 'Second Spring' is guided with a winding flow and separated layers. The sound is like wind and storm. People can sit and lie here all day, just feeling like inside deep mountains."[54] His words confirm that "The water of 'Second Spring'," as the source of the water scene, has always been a feature of the garden. But he also point to two fold appreciation of the water scene: visual and aural. Both, as we have established, were strengthened after the reconstruction.

The description of "winding flow and separated layers" is about the visual form and can be further understood through two distinct aspects. One is the "winding flow," a horizontal aspect of form that turns, similar to the previously existing scene "Winding Stream." But the description of "meandering flow in the rocky sunken ground in the mountain" by Li Nianci in 1686 shows the deep and serene feeling in the space surrounded by rocky cliffs, which is very different from that of a normal stream.[55] The other is the "separated layers," the vertical aspect of the form. The scene after reconstruction was also called "Three-layer Spring,"[56] which can be seen in poems by Wu Qi[57] and Qin Daoran.[58] In contrast to the name "Winding Stream," "Three-layer Spring" highlights the vertical aspect over the horizontal variation. Thus, we can derive that a new feature has been created by the construction of "separated layers."

The description of the aural aspect "like wind and storm," when compared with the "gurgling" in prior descriptions—tells us that it was now more overtly like that of a gully, where the sound of the water could be augmented and amplified. In Qian

Surun's verse "the flying spring sprays and makes sounds, like waves of tens of thousands of *qing* in the one gully."[59] Here we find the effect of the sound exaggerated even to a comparison with sea waves. And in Shao Changheng's *Record of Visiting Qin's Garden in Hui Mountain* reads, "the spring flows in the rock crevices with low and loud sounds. The low sound is near quietness while the loud one is more rushing. And when the water falls into the pond, the sound is like roaring."[60] Here we have various sound effects achieved by multiple manners of construction in different parts of the gully.

Perhaps the most important and wonderful effect of all is to be found in the "feeling like being in a ravine" (*ru-ru-yan-gu*). As mentioned above, "Zhang's Mountain" pays great attention to the bodily experience of the landscape. While the scenes of stream and waterfall in the late Ming were very attractive, they were nonetheless "views" to be appreciated outside, from specific points located in the "Pavilion of Hanging Flow" and the "Pavilion of Containing Greenness." These scenes were not "places" that could be traversed physically. In the new deeper gully with its high cliffs, visitors could actually pass inside and "sit and lie here all day," experiencing the realm of a ravine landscape—an experience that could never have been achieved before. This could also be an explanation of the effect described as "like natural creation," as concluded by Qin Songlin in the local chronicle of *Wuxi xian zhi*.

Perhaps now, the answer to why the large-scale reconstruction was necessary—why it was a marked improvement may become more apparent. Even though the previous scenes of the late Ming were already so compelling, the gully created by Zhang Shi substantially intensified the effects of the landscape features. With Zhang Shi's reconstruction, not only did the visual form become more coherent and naturalistic, but the garden landscape was made available to the senses, and could be experienced more fully in the soundscape and bodily experience. The special feature of "feeling like inside a ravine" had never been realised before. This gully, with the multiple sensorial experiences it offered, together with the main area of mountain and pond charged with the "painting-idea," has been the characteristic scene in Jichangyuan ever since.

Today, the "Gully of Eight Tones" in Jichangyuan differs slightly from the gully of the Qing period. The waterfall at the pond and the vertical division of "three layers" have nearly disappeared, while the soundscape is far from the descriptions we find in historic literature, "like wind and storm" or "like waves of tens of thousands of *qing*." This change could partly be attributed to the diminished flow of water from Second Spring. Yet another reason lies in the inferior renovations of the gully that took place during the 20th century. In *On Gardens*, Mr. Chen Congzhou, one of the most important scholars of Chinese gardens in modern times, expressed his regret that, "Deprived of its balance, the ravine has lost its former charm," and he asked, "Is it not clear that utmost discretion and meticulous care should be exercised in carrying out renovations of this kind?"[61] Despite this partial loss, when visiting today, as one moves along the gully of over thirty meters in total length, with cliffs of yellow stones—we still feel the secluded and serene realm. Surrounded by the flowing stream, rocky cliffs, and lush trees from below to above, the gurgling sound of water, and the spatial variation of narrowness and broadness while walking—all contribute to the pleasure of the experience of the deep mountain valley.

5 Conclusion

The reconstruction work of Jichangyuan in the 1660s was a significant success thanks to the great skill of Family Zhang—the leaders of the new direction in garden making in 17th century Jiangnan. The highly conducive and perhaps superior topographical condition of Jichangyuan contributed to its success as a masterpiece of garden making, which may even have surpassed the earlier project of Lejiao Garden that established Zhang Nanyuan's illustrious reputation. After combining a study of the current features of "Zhang's Mountain" with our comparison of the garden between the late Ming period and modern times, two outstanding achievements of the landscape reconstruction of Jichangyuan in the early Qing can be discerned. The first is the visual effect of the mountain foot with "painting-idea," and the second is the physical experience of visiting in a mountain ravine. These can still be clearly appreciated today after historical vicissitudes.

This successful reconstruction work of Jichangyuan also provides us with an ideal case for further understanding of"Zhang's Mountain" and the transformation that occurred in the aesthetic direction of garden-making in 17th century Jiangnan. The deep commitment and pursuit of the "painting-idea" in Zhang's new methods is apparent not only though visual appreciation of the landscape forms, but also in the corporeal experience so admirably realised both in the main area of "mountain-and-water" and the new winding gully. The reconstructed Jichangyuan is an outstanding masterpiece and the only remaining case of"Zhang's Mountain" in Jiangnan, which marks the high point of the art of 17th century garden making, and the peak of the Chinese garden history.

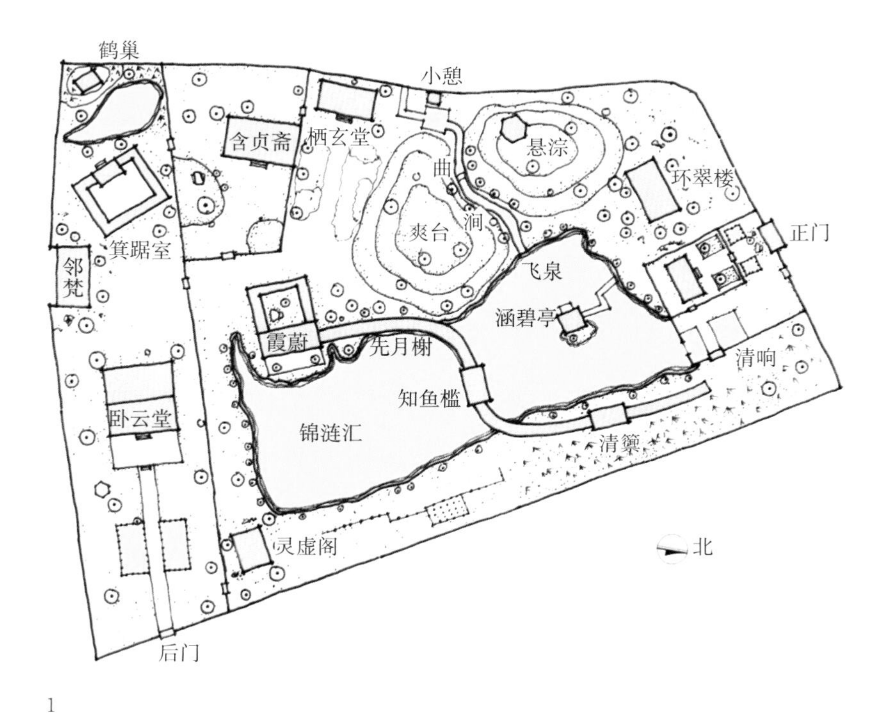

1

图 1：秦燿寄畅园平面复原示意图。黄晓绘制
Figure 1 : Conjectural plan of Jichangyuan in the 1590s. Drawn by Huang Xiao

图 2 : 宋懋晋《寄畅园五十景图》之“曲涧”
Figure 2 : “Winding Stream” in *Album of Jichangyuan* by Song Maojin

图 3 : 宋懋晋《寄畅园五十景图》之“知鱼槛”
Figure 3 : “Rail of Knowing Fish” in *Album of Jichangyuan* by Song Maojin

图 4 : 宋懋晋《寄畅园五十景图》之“飞泉”
Figure 4 : “Flying Spring” in *Album of Jichangyuan* by Song Maojin

图 5 : 宋懋晋《寄畅园五十景图》之“涵碧亭”
Figure 5 : “Pavilion of Containing Greenness” in *Album of Jichangyuan* by Song Maojin

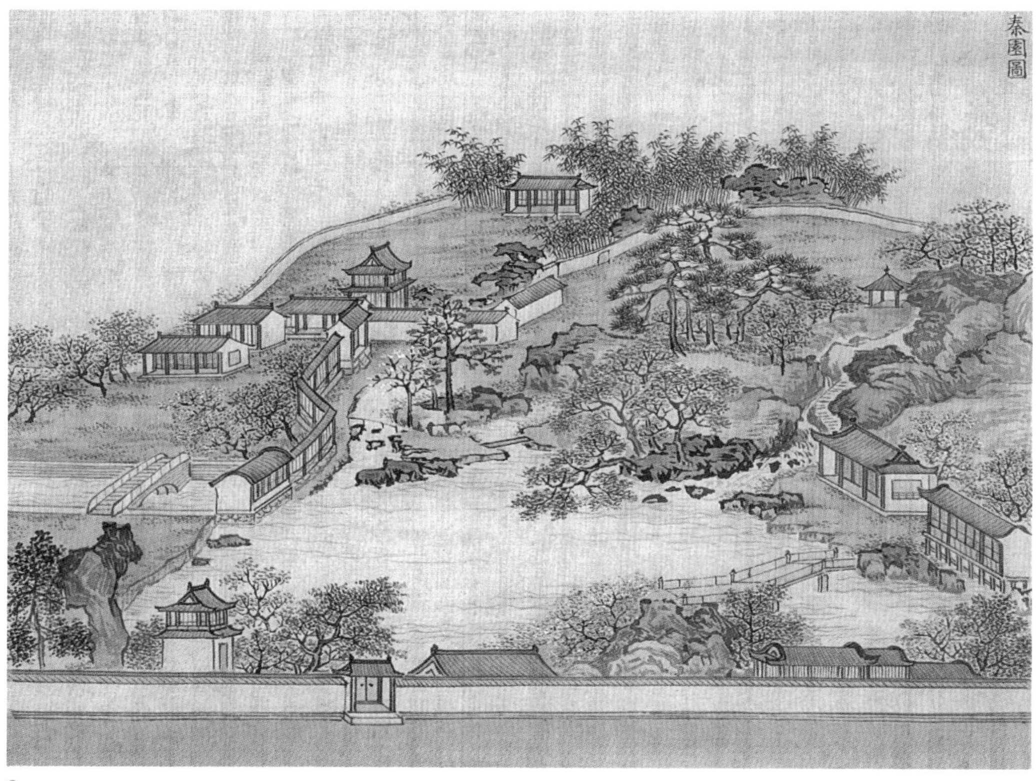

6

图 6 : 佚名《清高宗南巡名胜图》之“秦园图”
Figure 6 : “Painting of Qin’s Garden” in *Album of Scenic Spots of Qianlong Emperor’s Southern Inspection* by anonymous author

7

图 7 : 高晋《南巡盛典》之 “寄畅园”

Figure 7 : “Jichangyuan” in *Ceremonial Records of the Imperial Tours to the South* by Gao Jin

雨餘山滴
翠春暮卉
爭芳寥薜
盤雲逕拔
杉渡石梁
鳴湍空塵
意列岫滄
煙光更許
傳佳話遙
頒詩倩償
御製再題寄
暢園一首
臣梁詩正敬書

图 8：钱维城《弘历题寄畅园诗意卷》
Figure 8 : *Poetic Idea of Jichangyuan with Emperor Qianlong' s Inscription* by Qian Weicheng

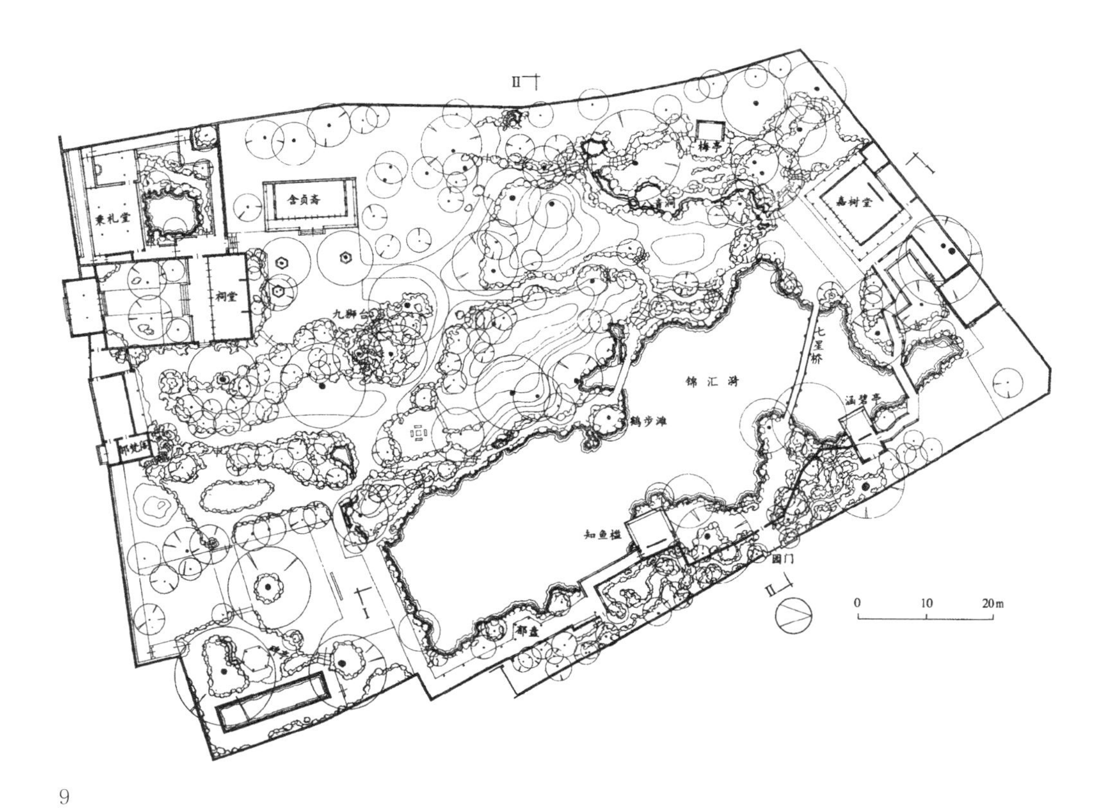

9

图 9：二十世纪八十年代寄畅园平面图。潘谷西编著，《江南理景艺术》（南京：东南大学出版社，2001 年），第 172 页

Figure 9 : Plan of Jichangyuan in the 1980s. In Pan Guxi, *Jiangnan lijing yishu* (*Landscape Art in Jiangnan*), (Nanjing: Southeast University Press, 2001), 172.

10

11

图 10 : 从“锦汇漪”东岸亭廊隔水西望。顾凯拍摄

Figure 10 : Scene seen from pavilions at the east bank of “Ripple of Brocade Converging”. Photo by Gu Kai

图 11 : 从“锦汇漪”北岸南望。顾凯拍摄

Figure 11 : Scene seen from the north bank of “Ripple of Brocade Converging”. Photo by Gu Kai

12

13

图 12：从“锦汇漪”南岸北望。顾凯拍摄

Figure 12 : Scene seen from the south bank of “Ripple of Brocade Converging”. Photo by Gu Kai

图 13：“八音涧”局部。顾凯拍摄

Figure 13 : Partial scene of “Gully of Eight Tone”. Photo by Gu Kai

Gardens Real and Imagined: Gardens, Landscape Painting, and *Biehaotu*

Alison HARDIE

This paper addresses the relationship between actual gardens, landscape paintings, and *biehaotu* (cognomen pictures, by-name pictures, studio-name pictures), with a particular focus on the last. Are these different things related, and if so, in what ways? Are biehaotu a kind of landscape or even topographical painting, or are they better understood as a kind of portrait? Do they depict in any way the actual gardens of their owners? It will become clear in the course of the paper that I have many more questions than answers, so this discussion is intended to highlight the questions, and I hope it will stimulate other scholars with a deeper knowledge of the topics to try to answer them.

As we are well aware, garden culture played a large part in elite life in the late Ming, and landscape painting was also an important genre of art from at least the Song Dynasty onwards. Biehaotu are perhaps less well known: they are defined as paintings of a landscape and/or garden named with the cognomen/studio-name/*biehao* of the recipient, apparently painted as a gift (although the idea of a simple "gift" is probably naïve in such a case; the work might be in some sense "commissioned" from the artist, or produced as part of an ongoing process of reciprocal social exchange).[1] an art-historical category, the genre was first identified by Zhang Chou (1577–1643), who saw them as particularly characteristic of Ming painting, that is the painting of his own time. After describing what he saw as the characteristic genres of previous Dynasties, and giving examples of these (which I have omitted here), he identifies biehaotu as the particular feature of Ming painting:

> In the Jin, narrative painting was most appreciated [examples omitted], in the Tang it was novelty [examples omitted], in the Song classical literature was depicted [examples omitted], in the Yuan sketches of pavilions and kiosks were produced [examples omitted], while the Ming has created biehao pictures, such as Tang Yin's "Maintaining ploughing" [Shougeng], Wen Bi's [Wen Zhengming] "Chrysanthemum Nursery" [Jupu] and "Bottle Hill" [Pingshan], and Qiu Ying's "Eastern Grove" [Donglin] and "Jade Peak" [Yufeng]. All these five types are worthy of appreciative attention, although the pavilions and kiosks [of the Yuan] are the most elegant.[2]

The genre was particularly associated with the artists of what is now known as the Suzhou School (*Wumen pai*); Wen Zhengming (1470–1559) and Qiu Ying (ca. 1494–ca. 1552), referred to by Zhang Chou, were two of the leading lights of that somewhat amorphous grouping. The genre of biehaotu began in the mid-Ming (the fifteenth century) and lasted until the early Qing, when it faded out. The prominence of the genre at this historical period is clearly connected with the importance of individuality and authenticity in the late Ming, a time when self-expression in literature, art, gardens and life was held in high importance by the elite. These literati chose names for their gardens and for themselves, especially *hao*/biehao/studio-names, to reflect and express their personality and ideals. The flourishing of the genre of biehaotu also coincides with the late-Ming flourishing of garden culture, and with the first appearance of named garden designers, such as Zhang Nanyang (born ca. 1517), Ji Cheng (1582–ca. 1642), and Zhang Lian (Zhang Nanyuan, 1587–ca. 1671). The emergence of these successful artisans, or (as seems likely in the case of Ji Cheng) low-level literati obliged to earn their livings by practical means, suggests the importance of gaen culture in the life of the elite, as well as the relative social mobility of the late-Ming period, which gave rise to new ways of thinking about life and leisure, aptly reflected in the phenomena of garden culture.[3] The close connection between landscape and personal identity is reflected in a dictum of Wen Zhengming at the end of an essay on the *Jade Maiden Pool* (*Yunü tan*) garden of Shi Ji (1495–1571): "Landscapes become important through people, but people become important through landscapes."[4]

The first specialised modern study of biehaotu was an article published by Liu Jiu'an in 1990, in which he identifies thirty surviving paintings as biehaotu.[5] Other scholars have come up with different numbers (see table), Xu Ke in 2016 identifying fifty-nine paintings, almost twice Liu Jiu'an's total.[6]

The variation in numbers of biehaotu identified by different scholars suggests the instability of the term: what appears to one person as a landscape or garden painting may be seen by another as a biehaotu. In what follows, after giving some examples of what I see as "standard" and "non-standard" biehaotu, I will try to problematise the category, raising some questions about how biehaotu can be defined.

The painting which appears to be the earliest known biehaotu is only recognised as such because Zhang Chou tells us that it was. This is Du Qiong's (1396–1474) *Befriending the Pines*" (*Yousong tu*), now in the collection of the Palace Museum, Beijing.[7] It has no dedicatory inscription, and we might think it was simply a depiction of a landscape garden, if Zhang Chou did not tell us that "Master Dongyuan [Du Qiong] drew this painting for his brother-in-law Wei Yousong."[8] As it is an early example of the genre, we may speculate that the convention of including an inscription on the painting to record the recipient and/or his studio-name had not yet become established.

A later and more typical example of a biehaotu is a hand-scroll of 1529 by Wen Zhengming, also in the Palace Museum, called *Thatched Hall of the Luo [River] Plain* (*Luoyuan caotang tu*).[9] The inscription on the painting reads "On the fourth day of the seventh month of the *jichou* year of Jiajing, [Wen] Zhengming drew [this] painting of the Thatched Hall of the Luo [River] Plain". On a separate sheet is an essay, "Record of the Luo [River] Plain" (Luoyuan ji), also dated 1529, composed and written by Wen Zhengming, which explains that the family of the recipient of the painting, Bai Yue (1498–1551, formal name Zhenfu 貞夫), came originally from the Luoyang area, but had made their home in Suzhou; Bai named his studio in honour of his family's place of origin. Again, we would not necessarily know from the painting alone that this was a biehaotu, but the accompanying essay makes this clear.

As Zhang Chou's remarks on biehaotu suggest, Wen Zhengming was a quite prolific painter of biehaotu; indeed, he complained in a letter to his friend Wang Ao (1450–1524) that he was constantly being pestered for essays on people's biehao, and presumably for pictures too (in the letter he is discussing writing rather than visual art).[10] An example of a painting by Wen Zhengming which was known to Zhang Chou and which was quite possibly a biehaotu is *The Cassia Grove Studio* (*Conggui zhai tu*) in the Metropolitan Museum of Art in New York, datable to before 1534. (Figure 1) This was painted for a Mr. Zheng, with the formal name Zichong, who has not been identified. As Richard Edwards points out, the cassia is associated with examination success, particularly in the provincial examinations, which took place in the autumn, the season in which the cassia produces its fragrant flowers; the red leaves of the vine on the pine tree to the left also indicate autumn, and are matched by the red robe of the scholar, presumably Mr. Zheng himself, who can be seen, half-concealed, in the rear portion of the studio. Poems by various writers attached as colophons also refer to exam success. Edwards points out the portrait-like nature of the garden picture or biehaotu: "Here Wen is painting for a patron who wants a portrait, really no less than a personal portrait, for the garden is but an extension of the man."[11] But the strong emphasis in the painting on examination success—the inclusion of the cassia trees and the autumn leaves, as well as the name Cassia Grove itself—makes us wonder whether the studio, or at least its pictorial representation, has any basis in reality at all: the painting seems to be primarily an expression of good wishes for, or congratulations on, passing the exams.

A painting very different in appearance, but still classifiable as a biehaotu, is Tang Yin's *Wutong Mountain* (*Tongshan tu*), also in the Palace Museum.[12] This appears to be a pure landscape painting, with no garden or studio component. It shows an extensive body of water, with some mountains in the far distance, and on the left a cliff plunging into the water, with a gnarled tree extending from the cliff, and a small stream or waterfall flowing down beyond the tree. An accompanying quatrain clarifies that the body of water is the River Huai:

> I have heard that the water of the Huai emerges from Wutong Mountain;
> From of old it has produced worthies and sages.
> Now you, sir, you have learned from them indirectly;
> Your angling fish-hook must make use of one of its coves.

A signature follows, reading "Painting of Wutong Mountain by Tang Yin of Wumen [Suzhou]". It is thus clear that the painting was created for an acquaintance of Tang Yin's with the studio name Wutong Mountain (who this was is not known). The reference to the "angling fish-hook" in the last line of Tang Yin's quatrain indicates that "Master Wutong Mountain"—or Tang Yin on his behalf—represented himself as a carefree angler or hermit in the wilds. Thus the painting's depiction of an apparently re mote, unpeopled scene is appropriate to the self-representation of the recipient or "subject" of the painting.

At the same time, as Liu Jiu'an points out, a biehaotu is as much an expression of the artist's personality as of the recipient's.[13] In this sense, Tang Yin's painting is also a representation of his own self-idealisation as a hermit, worthy or sage. The same could be said of gardens: the garden is a representation of the garden designer's personality as well as that of the owner. In this connection we may recall Ji Cheng's deliberately ambiguous claim at the beginning of *The Craft of Gardens* that"seven-tenths [of the creation of a garden] is the master's," where "master" could equally mean the owner or the master craftsman who has created the garden.[14] We know also that literati of that time were very conscious of the different styles of garden designers, and aware that a craftsman might know better than the garden owner how his garden should be laid out, from the comment by Wu Weiye (1609–1671) in his biography of Zhang Lian (Nanyuan):

> If the owner was a connoisseur, Mr. Zhang would not be pushed around but would stay on site to supervise the construction; some might insist on having their own way and he would be forced to bend to pressure, but later someone would be

> sure to come along and say with a sigh, "I'm sure this wasn't Nanyuan's own design."[15]

Given that a biehaotu can be a form of self-expression both of the artist and of the recipient, one interesting question that comes to mind is: can an artist paint a biehaotu for himself, or does it always have to be for someone else? One work that could be regarded as a self-painted biehaotu, coming from near the end of the seventeenth century, and thus towards the end of the period during which biehaotu were commonly produced, is Wu Li's (1632–1718) painting of 1679 entitled *Whiling away the Summer in the Ink Well Thatched Hall* (*Mojing caotang xiaoxia tu*), now in the Metropolitan Museum of Art in New York (Figure 2). Wu Li's *hao* was "Ink Well Man of the Way" (Mojing daoren),[16] and the painting certainly in some way represents his studio, his studio-name, and indeed himself, as the figure reclining on a chair in the "Thatched Hall" itself, which is represented as being entirely roofed with thatch, and also possibly as the figure holding a fishing-rod on the nearer bank of the stream near the bridge. We could see this painting as an expanded self-portrait, in that it shows not only the artist but the environment (both building and landscape) with which he identifies himself.

Another question is: does a painting of someone's garden count as a biehaotu—or something equivalent to a biehaotu, a form of portrait (as with Wen Zhengming's *Cassia Grove* painting, discussed above)—even if the owner does not generally use the garden name as a biehao? An example of such a painting is Qian Gu's (1508–ca. 1578) depiction, in a handscroll now in the Palace Museum, of the garden of the Suzhou playwright Zhang Fengyi (1527–1613), the *Garden of Seeking One's Own Aims* (*Qiuzhiyuan*).[17] An explanation for the garden's name—and it does not really matter whether this was actually how the garden was given its name or was merely a later justification—is given by Zhang Fengyi's friend Wang Shizhen (1526–1590) in his "Record of the Garden for Seeking One's Own Aims" (Qiuzhiyuan ji). In Wang's account, Zhang Fengyi describes how modest his own garden is in comparison with the many more splendid gardens of Suzhou, but says that within the garden he is able to fulfil his (evidently modest) ambitions and achieve his aims. Modesty aside, the fact that Zhang Fengyi was the recipient in 1550 of Wen Zhengming's great painting *Old Cypress* (*Gu bo tu*), now in the Nelson-Atkins Museum of Art, and its many colophons by distinguished locals, shows how well integrated he was into Suzhou society, despite his (claimed) relative poverty.[18] Wang Shizhen concludes his account:

> When Master Wang heard this he sighed and said: "How fine is that which you seek, sir! Will you permit me to hear what exactly your aims are?" Boqi [Zhang Fengyi] smiled without replying. Master Wang said, after an interval, "I take your point!"[19]

The garden and its name are therefore very closely tied to Zhang Fengyi's personal ideals and identity, but there is no evidence that he used "Seeking One's Own Aims" as a studio-name. So to a great extent Qian Gu's painting represents Zhang in the way that a biehaotu would do, but without being a biehaotu in any strict sense.

Zhang Fengyi's friend Wang Shizhen, and the painter Qian Gu, provide us with another work which could be considered as a sort of biehaotu, though it does not conform to the standard for a biehaotu of being an individual, stand-alone painting. This is the first leaf of an album entitled *Record of a Journey* (*Jixing tuce*), painted by Qian in 1574 to commemorate Wang's journey from his home in Taicang to the capital, to take up office for the first time after the execution and posthumous rehabilitation of his father.[20] This first leaf is by far the most elaborate in the album (in the collection of the National Palace Museum, Taipei), which mostly shows sketched landscape scenes along the Grand Canal. An inscription on the painting names the garden as "Lesser Jetavana" (Xiao Qilin), Jetavana being the location where the Buddha preached the sermon recorded in the *Surangama Sutra*. Lesser Jetavana was the name of one part within Wang Shizhen's large garden, which as a whole was named the Yanzhou or Yanshan Garden (Yanzhouyuan, Yanshanyuan), alluding to a mountain in Daoist mythology. Wang Shizhen was known as the Hermit of Yanzhou (Yanzhou shanren), so if this album leaf is regarded as a depiction of the Yanzhou Garden as a whole (though it shows only part of it), then it can be considered as a biehaotu, especially as it shows not only the garden but also the building within the garden specifically named the Yanshan Hall

(Yanshan tang). This is the large building in the lower left–hand quarter of the painting; it is not labelled as such in Qian Gu's painting, but can be identified from Wang Shizhen's own very detailed description of his garden, the "Record of the Yanshan Garden" (Yanshanyuan ji).[21] It was this building which would be used for formal entertainment of guests within the garden, so anyone to whom Wang showed the album would presumably be aware of the building's identity.

In considering garden buildings bearing the studio name of the owner, we may look back to another album, whose whereabouts are now unknown, Wen Zhengming's album of scenes from the Artless Administrator's Garden (Zhuozhengyuan) of Wang Xianchen (*jinshi* 1493). Wang used the studio name Sophora Rain (Huaiyu), and one of the album leaves presents Wen's impression of the Sophora Rain Pavilion (Huaiyu ting), shown as an open-sided thatched hut among trees, beside a stream crossed by a small bridge; the figure seated in the pavilion is presumably intended to represent the owner Wang Xianchen.[22] According to Wen's *Record* of the garden, at the very start of which he introduces its owner as "Master Sophora Rain":

> Below [an ancient sophora tree] one crosses the water by a footbridge; as one crosses this going eastward, shady dwarf-bamboo spreads around, elms and sophoras give shelter, and there is a spreading pavilion overlooking the water, which is the Sophora Rain Pavilion.[23]

The preface to the poem accompanying the album leaf is even more explicit in making the link between the building and Wang Xianchen's cognomen:

> The Sophora Rain Pavilion is to the south of the Peach Blossom Shore, and to the west it overlooks the Bamboo Burn. There are various different trees planted around: elm, sophora, bamboo, and cypress. "Sophora Rain" is Mr Wang's cognomen.[24]

An album of this sort—including Wen's Record, written out in his own calligraphy—is in a sense a sort of extended biehaotu for the recipient, relating his garden dwelling to his ideals as a person, in the same way as the studio-name encapsulates a person's ideals and aspirations. In this way it serves as an extended portrait, though in a different way from the extended self-portrait in the horizontal scroll by Wu Li discussed above.

Thinking about how a garden feature or studio building can reflect the personality of its owner prompts a further question about biehaotu: what is the relationship between the studio or other garden building or feature depicted in a biehaotu and the owner's actual studio building? Can we assume that the biehaotu depicts the building with a degree of verisimilitude, or is it quite fanciful? More broadly, what is the relationship between the garden or landscape environment shown in the biehaotu and the actual environment? If we consider Wen Zhengming's 1549 painting of *True Appreciation Studio* (*Zhenshang zhai tu*), in the collection of the Shanghai Museum, which was painted, as Wen's own inscription on the painting tells us, for the Wuxi art collector Hua Xia (ca. 1498–?), we can see that it is most unlikely that Hua Xia's actual studio was a thatched cottage in the deepest countryside, as suggested in the painting.[25] From a purely practical point of view, this would not be a secure place to store a valuable art collection, which Wen tells us, in the preface to his "dedication" (*ming*) for the Studio, included calligraphy dating back to the Wei, Jin, and Tang Dynasties.[26]

On the other hand, if we look at the painting by Wen Zhengming's son Wen Jia (1501–1583) of *Wang Baigu's Half-Gatha Hermitage* (*Wang Baigu Banjie an tu*), dated 1573 and now in the Palace Museum, Beijing, we can see a much more realistically plausible urban or suburban space.[27] Wang Baigu is Wang Zhideng (1535–1612), from Suzhou, who used Master of the Half-Gatha (Banjie zhuren) among many other studio-names. Although it is plausible, however, this is not to say that it is an optically truthful representation of the studio. To take another example from the work of Wen Zhengming, the studio shown in a painting dated 1543 entitled *Living Aloft* (*Louju tu*), which looks rather like that of Wang Zhideng in Wen Jia's painting, definitely did not even exist at the time of painting; we know from Wen's inscription on the painting that the owner, Liu Lin (1474–1561), intended to erect a building of more than one storey (a *lou*), but at the time of painting it had not been built:

> Although the building is not yet completed, in anticipation of it I have composed a poem and sketched its concept.[28]

This is a reminder that we should be very cautious in assuming that any building depicted in a painting represents an existing building, even if we know that a building of this name or belonging to this owner did exist at some point.

It is reasonable to suppose that there is a wide variation in the way that the studios of the recipients or "subjects" of biehaotu are depicted, depending on the preferences of the painters or the recipients themselves for a "realistic" or idealised scene: did the recipients, for example, wish to be reminded of their actual dwelling (perhaps because they were going away on official service) or did they wish to have a flattering representation of their gentlemanly status? Such a wide variation is also likely to be true of the depiction of gardens and how they relate to the wider landscape. Sometimes, congruence between a written and painted representation of the same garden (as in the case of Wang Shizhen's Yanshan Garden) suggests that the painting is a more or less "realistic" or optically faithful rendering of the actual garden. In other cases, we may simply not have any independent evidence for the garden's appearance, or we may suspect that the painting is an idealised version.

Another thing to bear in mind, of course, is that Chinese artists did not, by and large, practise painting *en plein air*. Their work might well be related to an actual scene, but it would be created from memory after their return to the studio. A good case-study of how different two artists' renderings of the same garden landscape could be is provided by two paintings, completed within about three years of each other, of the Caressing the Waters Manor (Fushui shanzhuang) in Changshu, belonging to the noted writer Qian Qianyi (1582–1664) and designed by the well-known garden designer Zhang Lian. The later of the two paintings, dated 1642, is by Qian's concubine Liu Rushi (1618–1664), and shows a largely flat landscape with, in the foreground, a rustic bridge over a stream, which seems to open out into a wider expanse of water, and a two-storey building in the middle distance, surrounded by trees.[29] According to a colophon written by Qian, Liu sketched (*tu* rather than *hua*) this scene on the spot, and the landscape shown in the painting certainly conforms to what we know of Zhang Lian's style of garden design. The earlier painting, dated 1639, and also in the Palace Museum, is by Bian Wenyu (fl. 17th century), from Suzhou, "a painter of small originality", according to James Cahill.[30] It does not seem to have been painted for Qian Qianyi, but for the painter's own amusement. His inscription on the painting reads:

> In the Lesser Spring [the tenth lunar month] of 1639 I visited the Caressing the Waters Manor at Yu Hill, just at the time when the maple woods were brightly coloured, resembling sunset vapours rising over Red Rampart [Mountain], so I drew this likeness of it, to enable recumbent wandering. Bian Wenyu.

"Recumbent wandering" of course refers to the famous words of the early painter Zong Bing (375–443); when no longer able to visit the mountains because of old age, he painted mountain landscapes on his walls to enable him to "travel while lying down". Bian's use of the phrase implies that he will keep the painting to remind him of the striking scene on his visit to the garden. The landscape looks entirely different from that painted by Liu Rushi. We can see that Bian, a student and adherent of the leading art critic Dong Qichang (1555–1636), has assimilated the actual landscape of the garden and its surroundings to the canonical Dong Qichang type of scenery: in his mind, this has become a more "truthful" representation of the garden than its mere physical appearance. The craggy mountain in the background presumably represents Yu Hill. It is worth noting that Qian Qianyi used the cognomen Old Commoner of Yu Hill (Yushan laomin), so if Bian had wanted to, he could have given this painting to Qian as a sort of biehaotu.

Finally, let us consider one other aspect of biehaotu, its relation to topographical painting. It seems unquestionable that there is a connection between biehaotu and topographical painting, the latter also closely associated, since the time of Shen Zhou (1427–1509), with the Suzhou School who were exponents of biehaotu. Topographical painting was not always intended to give an exact depiction of the actual landscape: such paintings of the Suzhou School "tended to be quite schematic in character,

... in all cases subjecting the scenery rigorously to the dictates of established style";[31] however, "the inclusion of well-known temples, shrines, bridges, and pagodas served to situate the viewer and also invested the paintings with meanings that transcended the mere representation of scenery by evoking the historical, literary, and religious associations of these monuments."[32] This transcendence of the mere representation of scenery is also a characteristic of biehaotu, since even when it may be the case that the garden or studio building is being represented with some degree of optical faithfulness, it always has a deeper meaning by being associated with the owner's personality and ideals.

One painting where we may see the possibility of an overlap in both style and content between topographical painting and biehaotu is *Boating in Autumn at Crane Islet* (*Hezhou qiu fan*), of 1653, by Xiang Shengmo (1597–1658), now in the Palace Museum.[33] Xiang is not strictly a member of the Wu School, since his family was from Jiaxing, which is in Zhejiang, but as the grandson of the great art collector Xiang Yuanbian (1525–1590) he was well connected with the art world of Suzhou and indeed modelled his style on that of Wen Zhengming. This painting certainly recalls the Wu School topographical style, as exemplified in the work of Qian Gu, such as his hanging scroll showing the approach to *Tiger Hill in Suzhou* (*Huqiu qian shan tu*).[34] Crane Islet was, as Xiang explains in his inscription on the painting, the garden of his friend Zhu Maoshi (zi Kuishi, fl. 17th century), and Xiang's inscription emphasises its historical origins as the garden of Pei Xiu (791–864) in the Tang Dynasty; as it happens, this garden, like that of Qian Qianyi, is also recorded as the work of the garden designer Zhang Lian.[35] It is not clear whether Xiang Shengmo intended this painting for Zhu Maoshi or for himself (though his inscription says that it was intended to record what a good time they had had), and in any case Zhu Maoshi does not appear to have used Hezhou or Fanghezhou as a biehao. So the painting is not a biehaotu as such, but none the less has some of the characteristics of a biehaotu, since it shows a (probably idealised) representation of the garden dwelling of a friend of the artist.

At the same time the painting bears, as we have noted, the hallmarks of a topographical painting. It is interesting that the angle from which the view is shown is similar to that in *The Holistic*

View of the Zhi Garden in Zhang Hong's album,[36] which James Cahill convincingly argued was influenced by the European visual culture introduced by the Jesuits in the late sixteenth and early seventeenth centuries through such books as Braun and Hogenberg's *Civitates Orbis Terrarum (Cities of the World*, Köln, 1572–1616), Abraham Ortelius' *Theatrum Orbis Terrarum* (*Spectacle of the World*, Antwerp, 1579), and Jerome Nadal's *Evangelicae Historiae Imagines* (*Illustrations of the Gospel Story*, Antwerp, 1593). Indeed, Cahill has argued strongly for the influence of European images, and especially the treatment of colour in European art, on Xiang Shengmo himself.[37] We can see in this painting by Xiang some of the features which Cahill notes as prompted by the sight of European images: the diagonal or zig-zag recession into space, the tilted ground plane, the cutff buildings in the foreground, the washes of colour, and the representation of reflections in water. But Cahill is careful to note that Chinese artists do not necessarily adopt European techniques as a way of producing more "realistic" painting, but often as a way of making purely imaginary scenes more optically convincing.[38]

So having begun by thinking that this is a painting which gives a fairly optically truthful rendering of Zhu Maoshi's garden (and, like Liu Rushi's painting of the Caressing the Waters Manor, it does accord with what we know of Zhang Lian's style), I have come round to thinking that this is really more of a fantasy, an expression of the escapist experience of a visit to this garden, which allowed the Ming loyalist Xiang to forget for a short time the collapse of his world less than a decade before. As he says in his inscription: "In the enjoyment of woods and waters, nothing can surpass this."

I would like to suggest a few tentative conclusions. Firstly, I do not believe we can make a really sharp distinction between landscape or garden paintings, topographical paintings, and biehaotu. A given painting may, as I have suggested above, have elements of more than one of these categories. The category of biehaotu in itself, as the difference in numbers put forward by different art historians indicates, is an elastic one. Given the nature of biehaotu—or pictures having some of the characteristics of biehaotu—as "portraits" of the gardenor studio-owner, it would be interesting to examine them also in relation to more conventional figure portraits, but this is too large an enterprise to undertake here. Consideration of paintings as different as Wen Zhengming's *The Cassia Grove Studio* and Xiang Shengmo's *Boating in Autumn at Grane Islet* makes me think that we should be very cautious (more cautious than I have been in the past) about believing that artists in the Ming (at least with the exception of the remarkable Zhang Hong) were even attempting to represent their friends' or patrons' gardens with any degree of optical faithfulness, though I do still think that we can use some paintings—with due care—as guides to what gardens of the time may have looked like as regards style and layout. I would suggest, therefore, that as far as physical appearance is concerned, the connection between biehaotu and the actual gardens or studios of their recipients is a fairly loose one; indeed I am inclined to think that the more "standard" a biehaotu, the less likely it is to represent the garden or studio's physical appearance (as in the case of the True Appreciation Studio), while a painting that is not strictly a biehaotu (such as the "Garden for Seeking One's Own Aims") may be much closer to an optically truthful representation. At the same time, gardens and biehaotu have much in common, in that both are intended as expressions of the garden owner's or biehaotu recipient's personality and ideals. Fundamentally, the point of a biehaotu, as of a garden, is to represent the recipient's idealised identity: in the term used by Wen Zhengming in his inscription on *Living Aloft*, it is the "concept" which counts, not the actuality.

Table: *Biehaotu* identified by various scholars

Scholar	Date of article	Painters										Total
		Du Qiong	Xie Jin	Shen Zhou	Wen Zhengming	Tang Yin	Zhou Chen	Qiu Ying	Lu Zhi	Cheng Dalun	Wen Congjian	
Liu Jiu'an	1990	1		1	8	10	1	5	2	1	1	30
Zhao Xiaohua[39] (i)	2000					11						
Han Jinlei[40] (i)	2016					>7						
Xu Ke (ii)	2016a & b	2	1	1	28	15	2	6	2	1	1	59
Ye Yu[41] (iii)	2016				?							

(i) Zhao Xiaohua and Han Jinlei's articles are specifically about Tang Yin.
(ii) The total for Tang Yin includes one painting identified as a collaboration between Tang Yin and Qiu Ying.
(iii) Ye Yu's article is specifically about Wen Zhengming, and does not give a total.

真实的园林与想象的园林：园林、山水画、别号图

夏丽森

（顾凯、查婉滢、戴文翼、应天慧、董茹、刘玲玲 译）

本文探讨的是真实园林、山水画和别号图之间的关系，并聚焦于别号图。这些不同的事物有关联吗？如果有，是如何关联的呢？别号图是一种山水画或实景画吗？还是理解为一种肖像画更好呢？它们以某种方式描绘了其主人的真实园林吗？在论文的展开中可以清楚发现，我的问题要比答案多很多，所以这一讨论的目的在于彰显这些问题，我也希望能够激发对这些话题有更深入了解的其他学者来努力提出答案。

我们都知道，园林文化在晚明的精英生活中占有重要地位，而至少从宋代开始，山水画也成为一种重要的艺术形式。别号图也许不那么出名，它们被定义为以接受者的别号为名的山水画或是园林画，似乎是作为礼物而绘制（虽然在这种情况下，简单“礼物”的想法很可能是过于天真的；从某种意义上说，作品可能是艺术家被“委托”的，或者是作为持续社交进程的一部分而创作的）。[1] 作为一个艺术史范畴，该类别最初被张丑（1577—1643 年）提出，他认为这是他所处的明代的绘画所特有。在描述他所看到的前代绘画的特色类型、并举例说明之后（我在这里略过），他认定别号图是明代绘画的特色：

> 晋尚故实……唐饰新题……宋图经籍……元写轩亭……明制别号，如唐寅守耕，文璧菊圃、

> 瓶山，仇英东林、玉峰之类。五等皆可览观，
> 惟轩亭最为风雅。[2]

在张丑看来，该绘画类型与吴门画派艺术家尤其相关，文徵明（1470—1559 年）和仇英（约 1494—1552 年）是其领袖，不过当时此所谓画派尚未得到确定性认识。别号图始于明代中期（十五世纪），一直延续到清初而逐渐消失。在晚明，社会精英们高度重视文学、艺术、园林及生活中的自我表达，别号图在这一时期异军突起，与对个人性和真实性的强调有着明显的关联。这些文人为其园林、也为自己择名，尤其是“号”（别号、斋名），来反映和表达其个性和理想。别号图的兴盛与晚明园林文化的繁荣也正相合，此时首次出现了著名的园林设计师，如张南阳（约 1517 年—？）、计成（1582—约 1642 年）、张涟（张南垣，1587—约 1671 年）。这些匠师的成功或下层文人不得不以实践手段来谋生（可以计成为例）的现象，表明了晚明精英生活中园林文化的重要性和相对的社会流动性。由此，也产生了对于生活和休闲的新型思考方式，在园林文化的现象中多有呈现。[3] 这种景观和个人身份的密切关联，反映于文徵明为史际（1495—1571 年）所作《玉女潭山居记》结尾的断言：“地以人重，人亦以地而重。”[4]

对别号图的首个现代专门研究，是刘九庵在 1990 年发表的一篇论文，他鉴定出了 30 幅现存的别号图。[5] 其他学者提出了不同的数量（见表 1），许珂在 2016 年鉴定出 59 幅画作，几乎是刘九庵计数的两倍。[6]

不同学者所鉴定的别号图数量有异，这说明了这个术语的不稳定性：在一个人看来是山水画或园林画，在另一个人看来可能是别号图。下文中，在给出一些我认为“标准”和“非标准”的别号图例子后，我将就别号图如何定义提出一些问题。

已知最早的别号图应是杜琼（1396—1474 年）的《友松图》（现藏北京故宫博物院），因为这是张丑提出的。[7]

它没有题词，如果张丑没有告诉我们这是“东原先生为其姊丈魏友松写此图”，[8] 我们可能会认为它只是描绘一座山水园林。由于这是该类别的一个早期例子，我们可以推测，在绘画上加上题词来记录接受者别号的惯例尚未确立。

稍晚的一个更典型的别号图例子，是文徵明 1529 年所绘的手卷《洛原草堂图》，也收藏在故宫博物院，[9] 画上题词为“嘉靖己丑七月四日徵明写洛原草堂图”。文徵明还有一篇《洛原记》，也撰于1529年，提及画的接受者白悦（1498—1551 年，字贞夫），虽安家于苏州，但其家族来自洛阳地区，白以其郡望来名其书斋。在此，我们也并非仅凭这幅画就知道这是一幅别号图，而是通过所附记文得以清楚了解。

正如张丑针对别号图所谈及的，文徵明是一位非常高产的别号图画家。他确实曾在一封信中向友人王鏊（1450—1524 年）抱怨，时常有为他人撰别号文的邀约缠身，估计作别号图也是如此（在信中他讨论的是撰文而非作画）。[10] 为张丑所知的文徵明《丛桂斋图》，极有可能是别号图的一例，作于 1534 年之前，现藏纽约大都会美术馆（图 1）。此画是为一位名为郑子充（子充为其字）的男子所作，其人未详。正如艾瑞慈（Richard Edwards）所指出，桂树与考取功名有关，尤其是秋季乡试，其时正值桂花飘香。画面左侧松树上藤蔓的红叶同样表明了是秋季，这与画中文人的红袍相映，推测为郑氏本人，他半掩着身子出现在书斋后部。所附尾跋中有多首赠诗，也与科考成功有关。艾瑞慈指出了这幅园林画或别号图的肖像性质：“此画是文徵明为想要一幅肖像画的赞助人所绘，它正是一幅个人画像，这座园林不过是此人的延伸。”[11] 但这幅画对考取功名的强调——包括桂树、秋叶和“丛桂”这个名字——让我们不禁怀疑，画中的书斋，或至少画中形象，是否真的有现实依据：这幅画似乎很大程度上是在表达对顺利通过科考的祝愿或恭贺。

唐寅的《桐山图》是一幅形式迥异的画，但仍可归类为别号图，也藏于故宫博物院。[12] 它似乎是一幅纯粹的山水画，没有园林或书斋内容。画里展现了一片辽阔的水域，远处有一些山脉，左侧的悬崖直插入水中，其上伸出一棵盘根错节的树，树的后面有小溪或瀑布流下。所附的一首绝句说明了画中之水为淮河：

> 吾闻淮水出桐山，古来贤哲产其间。
> 君今自称亦私淑，渔钩须当借一湾。

诗后的落款为“吴门唐寅作桐山图”，很明显这幅画是唐寅为一位别号桐山的熟人所作（此人未详）。此诗末句提及的“渔钩”暗示了唐寅或“桐山主人”本身将主人比作一个身居野外的逍遥钓者或隐士。因此这幅画对一个看似辽远且无人的场景的描绘，符合画作接收者（即该画之“题”）的自我呈现。

同时，就像刘九庵所指出的，别号图既是接收者，也是艺术家的个性表达。[13] 从这个意义上说，唐寅的画作同时也是对他作为隐士、名士或贤者的理想化自我的表现。同样的说法也可以用于园林：园林同时是其设计者与所有者的个性表现。在此方面我们可以回顾《园冶》的开篇部分，计成特意含混地声称“七分主人”，[14] 这里的“主人”一词既可以指园林的所有者，又可以指造园的主持匠师。从吴伟业（1609—1671 年）在《张南垣传》中的评论也可以看出，当时的文人对园林设计师的不同风格有很清楚的认识，一个匠师可能比园林主人更懂得如何布置其园林：

> 主人解事者，君不受促迫，次第结构；其或任情自用，不得已骫骳曲折，后有过者，辄叹息曰：“此必非南垣意也。”[15]

既然别号图可以同时是艺术家和接受者的一种自我表达形式，于是出现一个有趣的问题：艺术家能否为自己画别号图，还是必定要为别人画？有一幅别号图，是可以被认作自画的，绘于十七世纪末，也就是别号图创作的末期，即吴历（1632—1718 年）于 1679 年所作的《墨井草堂消夏图》，现藏于纽约大都会美术馆（图2）。吴历的号是“墨井道人”，[16]这幅画一定在某种程度上再现了他的书斋、别号甚至是他自己，如那位在茅草铺顶的“草堂”内斜倚于椅上之人，也有可能是那位坐在桥边河岸上手持钓竿之人。我们可以把这幅画视为一幅扩展了的自画像，因为它不仅表达了艺术家本人，而且还展现了用以识别他的环境（包括建筑和景观）。

另一个问题是：一幅描绘某人园林的画，是不是就能算作别号图或者相当于别号图的一种肖像画（就像前述文徵明《丛桂斋图》那样）——即便主人并不以园名为别号？钱穀（1508—约1578年）的手卷《求志园图》（现藏于故宫博物院）就是这类画作的一个例子，描绘了苏州剧作家张凤翼（1527—1613年）的园林。[17]张凤翼的朋友王世贞（1526—1590年）在《求志园记》中对园林的名称作了解释，至于这是否真的是此园得名的来由还是后来的解释，都无关紧要。根据王世贞的记述，张凤翼形容自己的园林与苏州许多更富丽堂皇的园林相比是多么质朴无华，但他表示，他能够在园林中实现自己（显然很朴素）的抱负，达到自己的志向。撇开质朴不谈，文徵明曾于1550年赠予张凤翼一幅《古柏图》（现藏纳尔逊-阿特金斯艺术博物馆），其上诸多当地名流的题跋都表明了他在苏州文人圈中是重要成员，尽管他（声称）自己相对贫穷[18]。王世贞在记文末尾写道：

> 王子闻之，叹曰：“善乎！子之求也。志则可与闻乎？”伯起笑而不答。王子有间曰：“命之矣。”[19]

因此，这座园林及其名称与张凤翼的个人理想和身份密切相关，但没有证据表明他用“求志”作为斋名。钱穀的画作在很大程度上是像别号图那样表现张凤翼的，但严格来说它并不是别号图。

张凤翼的朋友王世贞和画家钱穀为我们提供了另一幅或可被视为某种别号图的作品，尽管它不符合别号图作为一种单幅独立绘画的标准。这幅画是钱榖于1574年创作的《纪行图册》的第一开，[20] 图册记录了王世贞从太仓故里到京城的行程，这是王世贞在父亲被处死及平反之后的首次就职。此图册（藏于台北故宫博物院）主要展示了大运河沿线的风景摹写，其第一开最为精美。此画上的标题将这一园林命名为“小祇林”，在《楞严经》中记载祇林是佛陀布道之处。小祇林是王世贞的大园林中的一个部分，整座园林名为“弇州园”或“弇山园”，寓意道教神话中的一座山。王世贞自称为弇州山人，所以如果把这一册页视作对弇州园整体的描绘（虽然它只展示了其中一部分），那就可以视其为别号图；尤其是它展示的不仅是园林，还有园中特意命名为“弇山堂”的建筑。这个大体量建筑位于此画左下方，虽然钱榖的画中并没有标出，但从王世贞自己的《弇山园记》这一对此园的极详叙述中可以辨识出来。[21] 弇山堂是园林中正式招待客人的场所，所以每个能获得王世贞展示这本图册的赏读者都会意识到这座厅堂的身份。

在考虑以主人别号命名的园林建筑时，我们可以回顾一下另一本目前下落不明的图册，即文徵明为王献臣（1493年进士）所绘的拙政园景图册。王献臣的别号是“槐雨”，文徵明册页中的一幅即是“槐雨亭”，展现的是树间的一座开敞茅屋，旁边是小溪，其上横跨一座小桥；一人坐于亭中，应即是代表主人王献臣。[22] 文徵明有《拙政园记》（该记的首句就把园主人称为“槐雨先生”），其中写道：

> 其下跨水为杠。逾杠而东，篁竹阴翳，榆槐蔽亏，有亭翼然而临水上者，槐雨亭也。[23]

此册页所附诗歌的前言更加明确地说明了这座建筑与王献臣别名之间的联系：

> 槐雨亭在桃花沜之南，西临竹涧，榆槐竹柏，所植非一。云槐雨者，著君所自号也。[24]

这类图册——包括文徵明亲笔所书的《拙政园记》——在某种意义上是对园主人别号图的一种延伸，把他的园林居所与他个人的理想联系起来，就像别号概括了一个人的理想和抱负一样。如此，它就成了一幅延伸的肖像画，虽然与前述吴历那幅水平卷轴的延伸自画像在方式上有所不同。

当我们在思考一个园林或书斋如何能反映其所有者的个性时，会引发关于别号图的另一个问题：别号图中所描绘的书斋，或其他园林建筑或特征，与所有者的实际书斋之间的关系是什么？我们能假设别号图对其中建筑的描绘有一定准确性、抑或很大程度上是想象的吗？更宽泛地说，别号图所展示的园林或景观环境与实际环境之间的关系是什么？如果我们看一下文徵明 1549 年所作的《真赏斋图》（藏于上海博物馆，由文徵明在图上的题词可知是为无锡艺术收藏家华夏（约 1498 年—？）所作），[25] 就会意识到，华夏的书斋不太可能像画中所描绘的那样是在乡村最深处的茅草屋。从纯粹实用的角度来看，这不是存放有价值的艺术收藏品的安全之所，而文徵明在他为书斋所作“铭”的前言中告诉我们，藏品中包括魏、晋、唐时期的书法。[26]

另一方面，如果我们看一看文徵明的儿子文嘉（1501—1583 年）于 1573 年所作、现藏于北京故宫博物院的《王百谷半偈庵图》（百谷名稚登（1535—1612 年），苏州人，号半偈主人），我们可以看到一个更加现实可信的城内或城郊空间。[27] 虽然这看似可信，但并非意味着就是在完全忠实地再现这一建筑。再以文徵明的另一幅作品《楼居图》（作于 1543 年）为例，其中的建筑与文嘉笔下王稚登的书斋很像，但这在当时绝对不存在。我们从文徵明的题词中得知，主人刘麟（1474—1561 年）打算建造一座楼，但在作画时它还没有建成：

> 楼虽未成，余赋一诗并写其意以先之。[28]

这提醒我们，即使我们知道在某个时候确实存在一座有该名的或属于该主人的建筑，也应非常谨慎地去假定画中所绘的是该实存建筑的再现。

可以合理假定，根据画家或接受者自己对于“写实”或理想化场景的偏好，对别号图接受者或“主体”的书斋的描绘方式会有很大的差异。例如，接受者是希望能记起他们的实际住所呢（也许是因为他们要去外地做官），还是希望他们的士绅身份得到恭敬地体现呢？这种广泛的差异性也可能出现在描绘园林以及相关的更广阔风景的时候。有时，对同一园林的文字和绘画表达的一致性（如王世贞弇山园的情况）表明，这幅画或多或少是对实际园林的“写实”或视觉上的可靠描绘。在其他情况下，我们对园林的面貌可能根本没有任何独立的证据，或者我们可以怀疑这幅画是一个理想化的版本。

当然，另一件要记住的事情是，中国艺术家基本上没有现场对景作画的。他们的作品可能与一个真实的场景有关，但那是他们回到画室后凭记忆的创作。有一个很好的案例，是两位艺术家对同一园林景观的不同描绘，此园是著名作家钱谦益（1582—1664 年）的常熟拂水山庄，由著名造园家张涟设计，而两幅画作的完成时间相差约三年以内。其中较晚的一幅是 1642 年钱谦益之妾柳如是（1618—1664 年）的作品，画中的景色大致平坦，前景是一座横跨溪流的乡村小桥，小溪看起来像是通向更广阔的水域，中景是一座两层建筑，以树环绕。[29] 根据钱谦益的题跋可知，柳如是在现场勾勒了这一场景（“图”而非“画”），而图中的景观也与我们所知的张涟的造园风格是一致的。那幅更早一点的画作，是 1639 年由苏州的卞文瑜（活跃于十七世纪）所作，现藏于故宫博物院。据高居翰（James Cahill）所说，卞文瑜是一位“创意有限的画家”。[30] 这幅画似乎不是为钱谦益画的，而是为画家自己消遣而作。他在画上的题词是：

> 己卯小春，过虞山拂水山庄，正枫林如染，
> 大类赤城霞起，遂仿佛图此，以当卧游。
> 卞文瑜。

“卧游”当然是指早期画家宗炳（375—443年）的名言，当他年事已高不能去游山时，便把山水画在墙上，以便“躺卧游览”。卞文瑜使用该词，暗示他将保留这幅画，以让他回忆起参观此园时所见的迷人景致。这幅风景看起来和柳如是画的完全不同，我们可以看到，卞文瑜作为著名艺术评论家董其昌（1555—1636年）的学生和追随者，已将园林及其周围的实际风景吸纳入董其昌式的典型景观：在他看来，比起这一园林的实在样貌，这才成为更为“真实”的再现。背景中的陡峭高山推测是代表了虞山，值得一提的是，钱谦益自号“虞山老民”，所以如果卞文瑜愿意的话，他可以把这幅画作为一种别号图送给钱谦益。

最后，我们来考虑别号图的另一个方面，即它与实景画的关系。别号图与实景画之间的联系似乎是毋庸置疑的，从沈周（1427—1509年）的时代开始，实景画与作为别号图倡导者的苏州画派就有着密切的关联。实景画并非总是要准确地描绘实际的风景：苏州画派的这类画作“往往具有相当的图解性，……在任何情况下都严格地服从于既定风格的规定”；[31] 然而，“著名寺庙、圣祠、桥梁和宝塔的加入，让观者置身其中，并通过唤起这些名胜古迹的历史、文学和宗教联想，赋予了这些画作超越纯粹景物表象的意义。”[32] 这种对纯粹景物再现的超越也是别号图的一个特点，因为即使对园林或书斋的再现在某种程度上有着视觉真实性，也总是关联着主人的个性与理想而具有更深刻的含义。

我们可以看到实景画和别号图在风格和内容上都可能有重叠的一幅画，是项圣谟（1597—1658年）于1653年所作的《鹤洲秋汎》，现藏于故宫博物院。[33] 项圣谟家住浙江嘉兴，不是吴门派的正式成员，但作为大艺术收藏家项元汴（1525—1590年）的孙子，他与苏州的艺术圈有着密切联系，他的艺术风格也确实效仿了文徵明。这幅画无疑让人想起了吴门画派的写景风格，比如钱穀的立轴画《虎丘前山图》。[34] 按项圣谟在这幅画的题词中所说，鹤洲是他的朋友朱茂时（字葵石，活跃于17世纪）的园林，并强调其历史渊源是唐代裴休（791—864 年）的园林；恰巧，和钱谦益的园林一样，这座园林也被记载为造园家张涟的作品。[35] 尚不清楚项圣谟是意在为朱茂时还是为自己画了这幅画（尽管他的题词说这是为了“以纪其胜”），无论如何，朱茂时似乎没有把“鹤洲”或“放鹤洲”用作别号。因此，这幅画本身并不是别号图，但具有别号图的一些特征，因为它展示了这位画家的朋友的（很可能理想化的）园居形象。

与此同时，正如我们已经注意到的，这幅画具有实景画的特征。有趣的是，景物呈现的视角与张宏《止园图册》中的全景视角相似，[36] 高居翰令人信服地指出，这一视角受到了 16 世纪末 17 世纪初的耶稣会士所引入的欧洲视觉文化的影响，比如布劳恩（Braun）和霍根贝格（Hogenberg）的《全球城色》（科隆，1572—1616 年）、亚伯拉罕・奥特利乌斯（Abraham Ortelius）的《全球史事舆图》（安特卫普，1579 年）和杰罗姆・纳达尔（Jerome Nadal）的《福音史事图解》（安特卫普，1593 年）等书。事实上，高居翰还强有力地指出了欧洲图像，尤其是欧洲艺术中的色彩处理对项圣谟本人的影响 [37]。从项圣谟的这幅画中，我们可以发现高居翰所注意到的一些因见到欧洲图像而产生的特征：对角线或之字形的空间后退、倾斜的地面、前景中裁切的建筑、色彩的晕染以及水面的倒影。但是高居翰谨慎地指出，中国画家采用欧洲的技法并不一定是为了创作更为“现实主义的”绘画，而往往是为了让纯粹想象的场景更具视觉真实性。[38]

因此，我一开始以为这幅画对朱茂时的园林进行了视觉上相当真实的描绘(并且就像柳如是对拂水山庄的画作一样，它确实符合我们所知的张涟风格），后来我则认为这更多的是一种幻象，表达的是游此园时逃避现实的体验，这让忠于明朝的项圣谟短暂地忘记了近十年前他那个世界的崩溃。正如他在题词中所说：“林泉之乐，不过是矣。”

我想提出几个初步的结论。首先我不相信我们能把山水画、园林画、实景画和别号图明确区分开来。如前所述，一幅画可能包含多个这些类别的元素。由不同艺术史学家提出的画作数量的差异表明，“别号图”范畴本身就是有弹性的。鉴于别号图（或具有某些别号图特征的绘画）的特性是园林或书斋主人的“画像”，将其与更传统的人像图进行关联考察会很有意思，但这是一个太庞大的项目而无法在此进行。考虑到与文徵明的《丛桂斋图》和项圣谟的《鹤洲秋汎》等不同的绘画作品，我认为我们应该非常谨慎（比我个人以往更加谨慎），不可轻易相信明代的艺术家（至少除了不同寻常的张宏以外）会试图以某种视觉真实来再现他们的朋友或赞助人的园林，尽管我仍然认为我们可以利用一些绘画——在保持应有审慎的前提下——来引导我们对当时园林的风格和布局的认识指引。因此，我认为就外在形象而言，别号图与其接受者的实际园林或书斋之间的联系是相当松散的；事实上，我倾向于认为，越是“标准”的别号图就越不可能再现园林或书斋的外观（如“真赏斋”案例所示），而严格意义上不是别号图的一张画作（如“求志园”）则可能更接近视觉真实的再现。与此同时，园林与别号图有许多共同之处，都意在表达园林主人或别号图接受者的个性和理想。从根本上说，别号图的意义如同园林，是为了代表接受者的理想化身份：用文徵明在《楼居图》题词中的话来说，重要的是“意”，而非现实。

表 1：不同学者所认定的别号图

学者	发表时间	画家										共计
		杜琼	谢缙	沈周	文徵明	唐寅	周臣	仇英	陆治	程大伦	文从简	
刘九庵	1990 年	1		1	8	10	1	5	2	1	1	30
赵晓华[39] (i)	2000 年					11						
韩金磊[40] (i)	2016 年					>7						
许珂 (ii)	2016 年 a 和 b	2	1	1	28	15	2	6	2	1	1	59
叶玉[41] (iii)	2016 年				?							

(i) 赵晓华、韩金磊的文章都是专门写唐寅的。
(ii) 唐寅的统计包括一幅唐寅与仇英合作的画作。
(iii) 叶玉的文章专门写文徵明，并没有给出总数。

1

Figure 1 : Wen Zhengming, *The Cassia Grove Studio,* ca. 1532, handscroll, ink and colour on paper, image 31.6 × 56.2 cm. The Metropolitan Museum of Art. Public-domain image. https://www.metmuseum.org/art/collection/search/44601 (accessed 23 October 2018)
图 1：文徵明《丛桂斋图》，约 1532 年，手卷，纸本水墨设色，画幅 31.6 厘米 × 56.2 厘米。大都会艺术博物馆，公共领域图片。网址：https://www.metmuseum.org/art/collection/search/44601（2018 年 10 月 23 日访问）

2

Figure 2 : Wu Li, *Whiling away the Summer in the Ink Well Thatched Hall*, 1679, handscroll, ink on paper, image 36.4 × 268.6 cm, part. The Metropolitan Museum of Art. Public-domain image. https://www.metmuseum.org/art/collection/search/49158 (accessed 5 October 2018)
图 2 : 吴历《墨井草堂消夏图》，1679 年，手卷，纸本水墨设色，画幅 36.4 厘米 × 268.6 厘米，局部。大都会艺术博物馆，公共领域图片。网址：https://www.metmuseum.org/art/collection/search/49158（2018 年 10 月 5 日访问）

Hills and Ravines in the Heart: A World without "Space"

XU Yinong

In his 1957 essay, "Suzhou de yuanlin" (Gardens of Suzhou), Liu Dunzhen (1897–1968) employs the word *kongjian,* a modern Chinese equivalent to "space," in discussing issues concerning "spatial organization and assemblage" (*kongjian zuhe*) in gardens.[1] Limited as the employment of the word is in the essay, it nevertheless demonstrates a discursive orientation to be taken more intensively in his highly influential *Classical Gardens in Suzhou* (*Suzhou gudian yuanlin*), posthumously published in 1979, a work that not only contains a sub-section—"Scenic Areas and Space" under "Layout"—devoted to the concept of space, but is fundamentally threaded through with this concept.[2] The momentum was now set, and "space" thereafter has become a central, taken-for-granted idea in the studies of gardens by Chinese architectural historians, and in some scholarly approaches it has even been taken as a frame of reference, where discussions of garden are subsumed under this concept.

The trajectory of Liu Dunzhen's work runs in contrast to that of Tong Jun's (1900–1983). The latter scholar was of course not at all unfamiliar with the concept of "space,"[3] but his 1936 essay, "Chinese Gardens: Especially in Kiangsu and Chekiang," carries the word "space" only once, in a context largely meaning "area" or "ground,"[4] whereas his classic work, *Notes on the Gardens in Jiangnan* (*Jiangnan yuanlin zhi*), completed in 1937, presents a modern, albeit traditionalist, reading of gardens in the Jiangnan region without a single mentioning of "space" or any implication of this concept.[5] Interestingly, in an 1978 text in English, titled "Suzhou yuanlin" and intended to be an introduction to Liu Dunzhen's *Suzhou gudian yuanlin*, Tong Jun starts to write such sentences as that, in the garden, "curves and studied irregularity ... characterize the design, and space disposition limits visibility to one single pictorial courtyard" and that "care was taken to achieve contrast through open versus closed space, ..."[6]

The brief reference here to the works of the two dear teachers, both being recognized as forerunning modern architectural historians in China, is not intended for a comparative study of their individual approaches to garden studies,[7] nor does it imply that they are representative of all of the works on garden produced in their time. Rather, it is meant to serve as a starting point for a discussion of possible misinterpretation of Chinese gardens. It suggests that, in China, the use of the concept of "space" as a frame of reference in discussion of garden design is a modern phenomenon, that this phenomenon started from modern development of architecture,[8] and that "space" may not be the mode of perception or even a principal one. Chinese garden has been strongly "architectural," and the majority of academic studies of garden have taken place in architectural schools. It is therefore only natural that modern architectural concepts and terminologies such as "space" should find their way to the readings of gardens.

The situation seems more complex, however. Garden making and reading is inherently part of the humanities and thus part of theories of culture. As such, analysis and discussion of gardens is, to borrow Clifford Geertz's words, "not an experimental science in search of law but an interpretive one in search of meaning."[9] It follows then that its language and vocabulary have to be shared more or less among other separate areas of the humanities. This is especially the case in pre-modern China; in this "unitary civilization," as observed by F.W. Mote, "many of the key terms are employed in apparently parallel fashion, and the values present in one activity are readily convertible to those of others."[10] Thus the same issues one would raise, and similar terminologies one would employ, in the study of literary prose, poetry, painting, calligraphy and philosophy also apply to gardens. The bond between garden making and landscape painting is particularly tight: unequivocally two separate disciplines, they are closely interrelated and sometimes even inter-translated, each tradition frequently influencing and enriching the other, and this interrelatedness and inter-translatability allows for parallel readings of landscape painting and garden. Yet the application of the concept of space in studies of pre-modern landscape painting, as will be shown later in a few selected examples, can hardly be attributed to the influence of modern architecture, but to a more generic modern conception of the world.

The question I would like to raise in this essay is whether the modern injection of this overarching concept might be as much beneficial as detrimental to our perception of garden and landscape painting, in a way that, though invaluable in its own right, it has the tendency of excluding other interpretative possibilities and thereby hampering better understandings. Recent attempts in scholarship at challenging modern habitual ways of reading Chinese garden and landscape painting are encouraging,[11] but the shortcoming of these challenges lies in the fact that they operate entirely within the same conceptual framework, in which all of the concepts are interrelated, as the target of the challenges does; hence the effect of self-denial resulting in the notable limitation of argument.

* * *

Of the construction of "pictorial space"—a modern term[12] and a modern concept—in typical Chinese landscape paintings, a widespread interpretation is that viewing happens in "perspectives" with multiple or dispersed or shifting viewpoints, or simply multiple or dispersed or shifting perspectives; a diminishing in the size, of receding objects is clearly present, and so is atmospheric perspective with a decreasing in receding objects of contrast, details, and colour saturation; and parallel projections are often followed in the depiction of furniture, buildings and other architectonic structures. Such an interpretation, centred on "viewpoint," "perspective" and "projection," is certainly mooted by seemingly antithetical reference to studies of paintings in post-Renaissance Europe. Apart from technical drawings presented by a specific type of parallel projection, a viewing of European painting is, as generally believed, conditioned by a fixed vantage point for the whole picture in linear perspective, which creates an infinite, isotropic and homogeneous space as a consistent medium in which the depicted objects are located; that is, as postulated by Erwin Panofsky, "a fully 'perspectival' view of space."[13] Here, spatial consistency is crucial in perception. Perspective operates "as a generator *of* pictorial space," which "is applied to an entire picture not just to individual objects in the picture";[14] its most important function is, as William V. Dunning succinctly puts it,"to create a unified rational space."[15] The contrast between the two traditions appears to have become more salient after the introduction of linear perspective into China from the 17th century onward, as the two modes of perception are often seen as co-existing in the Chinese pictorial world.

Attractive and easy to comprehend as it is, this interpretative contrast reductively over-simplifies the European pictorial traditions.[16] Since this essay is not focused on the complexity of the concept of space, development of perspective or history of perception in Europe, suffice it here to say that scholarly debates throughout the past thirty years in the development of perspective in art and architecture have raised our awareness, and advanced our knowledge, of the questions of the historical framing of the "rediscovery of linear perspective" in Quattrocento Italy; of the conception of homogenous space, thought to be a necessary condition for the advent of linear perspective; and of the concept of space itself.[17] Nevertheless, more relevant to our discussion here is the realization that ideas of space are never static in European painting from the eighteenth century onward; Norman Bryson insists that one of the widespread but erroneous beliefs about space in European painting is that "'Quattrocento' space reigns unchallenged from Giotto [ca. 1267–1337] until Cézanne [1836–1906]," pointing to many historical cases, even just in France, of breaking with the Albertian conventions, including Antoine-Jean Gros (1771–1835), Jean-Auguste-Dominique Ingres (1780–1867), Ferdinand Victor Eugène Delacroix (1798–1863), and rococo painting.[18]

One cannot, however, deny the existence and application in post-Renaissance Europe of the highly influential concept of homogenous, systematic space and along with it of linear perspective that helps create it, even though this has been but one of many approaches to painting; hence the conceptual basis on which the interpretative contrast between China and Europe was constructed, and which functions in turn as a frame of reference for reading Chinese landscape painting. And yet on a closer look at the Chinese pictorial experience, one is tempted to suspect that this interpretative contrast may not be just an over-simplification. A suspicion like this, relating in particular to perspective, has already been signalled by quite a number of scholars, including Joseph Needham and James O. Caswell. Admirably sensitive to the intellectual, cultural and historical differences between China and Europe, Needham suggests that the Chinese convention of drawing (and painting) followed what he calls "the diffuse view-region principle," under which "there was a 'multiple station-point', or rather a 'hovering or dynamic view-region', not a viewpoint at all."[19] Needham's observation, albeit still operating within the frame of reference established on the basis of European pictorial tradition, is so acute as to verge on breaking away from it. Caswell, apparently taking for granted the same frame of reference in his statement that "... the Chinese were not ignorant of some use of perspective, but ... they developed their own approaches for their own purposes," tacitly and yet insightfully turns to a line of thinking similar to Needham's: "'perspective' is a rather inadequate term to be used for Chinese paintings as it carries the burden of Western conceptions."[20]

Both Needham's and Caswell's arguments are closing in on the crux of the matter, leading us to a questioning of the basic assumption from which the issues of both viewpoint and perspective have arisen—images presented in landscape painting have to be viewed from certain viewpoint(s), which should naturally bring about perspective of one sort or another. The fact is that, when images in landscape painting, such as those in *A Solitary Monastery amid Clearing Peaks* attributed to Li Cheng (919–ca. 967), are scrutinised, "multiple or dispersed or shifting viewpoints" do not really exist; that is, for their very existence, definite vantage points, no matter how much in multitude, how widely in dispersion and how frequently in shifting, each need to be precisely determined, but this is not the case in the painting. Thus, talking of "multiple or dispersed or shifting viewpoints" or even of "hovering or dynamic view-region" merely makes an approximation, edging toward the purview of linear perspective, so as for the discussions to remain logical and viable in the established frame of reference. In this sense, "perspective" as a term to be used for Chinese paintings is, with all respect to Caswell, more than just inadequate, it is irrelevant to them; the burden it carries is very heavy indeed, and it is misplaced as well.

The introduction of European representational modes and pictorial techniques, intensified from the eighteenth century onward, brought about notable changes to pictorial practice in China, as evinced in a number of the Qing court painters' work and in a large number of woodblock prints, where linear perspective is employed, more or less fully in some cases, but partially or superficially in some others.[21] It is precisely these cases of partiality and superficiality that afford me another opportunity to illustrate my point. And a similar situation is also found in *ukiyo-e*, a genre of Japanese art having flourished from the 17th through 19th centuries, which can also be referred to in support of my argument. Such references are justified, for a number of interrelated pictorial conventions, principally *yamato-e* and *kara-e*, both derived, either partly or fully, from Chinese painting traditions in different historical periods, had already been long established and were at the disposal of Japanese artists by the 1730s when some of them began to adopt linear perspective. Examples from China and Japan during this period are legion, but space limitation allows me to cite only a few works, first of two *ukiyo-e* artists—Okumura Masanobu (1686–1764) and Katsushika Hokusai (1760–1849), and then of the Qing artist Xu Yang (fl. 1751–ca. 1776).

* * *

Credited with a number of important technical inventions or developments, such as the *benizuri-e, hashira-e*, and *urushi-e*, Okumura Masanobu is also recognised for his instrumental role in popularizing the so-called perspective print known as *uki-e*.[22] One of Masanobu's well-known prints is entitled *Enjoying the Cool of the Evening at Ryôgoku Bridge, an Original Perspective Print* (*Ryôgoku-bashi yûsuzumi uki-e kongen*), created in about 1745 (Figure 1). In this picture, the interiors of the building appear to be in perspective, whereas the exterior scenes are not. Timon Screech holds that "Japanese architecture ... in some ways ideally suited to essays in perspective" in that the columns and beams of timber frame structure provided guide lines and woven tatami mats served as markers for easily plotting recessions, and yet "outdoor space remained more of a challenge, though."[23] In a similar line of thinking, Stephen Little's remarks are more unambiguous:

> Masanobu created numerous outdoor scenes utilizing his understanding of both singleand multiple-point perspective. These experiments often revealed basic misconceptions owing to his incomplete understanding of both the rules of perspective and the ontological assumptions that underlay them in their original European context.[24]

Ontologically, the assumptions on which the images were conceived by Masanobu may well have been very different indeed from those in the European contexts, but I am not very convinced that the obvious disparity between inside and outside as presented in the picture should be interpreted as demonstrating an inability or a lack of understanding on Masanobu's part. First, the perspective of interiors of this picture is only roughly suggested, as only three sets of lines supposed to be perpendicular to the pictorial plane converge at a single vanishing point, while others do not. Secondly, in his *Elegant Female Version of Shutendoji, an Original Perspective Print* (*Fûga onna shutendôji, uki-e kongen*), (Figure 2) probably created in the same year, the"spatial" disparity between the interior and exterior is equally significant, but in a totally reversed way—the scenery outside of the build–ing is rendered in a fashion that is "spatially" quite logical by the standard of European perspectival convention, whereas the inside, clearly suggestive of the use of perspective, is in disarray. It does not make much sense, therefore, to regard Masanobu as being capable or incapable of correctly applying linear perspective to his images, whether it be interior or exterior.

An alternative explanation would be that there were a wide range of options available to *ukiyo-e* artists, and to Masanobu. Not only did he selectively employ different conventions for different parts of a single picture, but also freely made adjustment to the rules of any single convention, so as to maximize the intended messages that the images convey. In the instance of the first picture, there are two main scenarios of activity inside the building, one in the centre in the foreground, and the other in the adjacent compartment on the right. It is possible to interpret the perspectival inconsistency that the lines formed by the lintels separating the central and right compartments are tilted upwards in different degrees, so that the ceilings of the right compartment can be exposed as much as possible, but the lines formed by the lintel on the right edge of the right compartment are tilted downwards, which to some extent helps close off this area, thereby turning the river image in distance into a backdrop. On the outside, even though the bridge on the left of the picture, the boats under the bridge and the boats near the centre of the picture together constitute a middle ground, the entire outdoor area is treated as one totality, thus functioning as a backdrop in terms both of activity and of image; the tilting up of river, bridge and open grounds may well have been intended to show clearly all the outdoor activities on the scene—boat-rowing, bridge-crossing, restaurant-seating, and so forth—which are set in contrast to the indoor leisurely and relaxing atmosphere; the latter is a world of its own, separated from the busy daily life in the outside.

This interpretation seems happily to resonate with Screech's acute observation:

> Perspective was enlisted to show precisely what was *not* quotidian, or vouchsafed, but the mimetic space of the stage, the "otherness" of the recreational or escapist worlds, or the pleasure zones beyond the actual.
>
> ... *Uki-e* were in cahoots with boastfulness and swell. They could be incorporated into narrative contexts; many is the illustrated story which runs comfortably along using traditional spatial configurations, only to switch into perspective (often applied *ad hoc*) when a flashy or insolent space is invoked.[25]

The ontological function of perspective in the *ukiyo-e* pictures, as argued by Screech, is thus a reverse of that of European paintings. An appreciation of this reversal of the function of perspective helps support my argument that linear perspective was taken as a new, alien devise, appropriated to work with traditions of what are known as Song-Yuan and *yamato-e* modes of pictorial presentation.

In the *Elegant Female Version of Shutendoji*, however, the indoor activity forming the foreground is concentrated in the nearest compartment only. To provide plenty of room and a good *mise-en-scène* for it, the lintels and ceiling of this three-bay compartment plus the area of one more bay further in distance are set in a "perspective" dramatically different from those of the compartments beyond. The area closer to the viewer, where all of the figures in foreground are grouped, is kept not only wide, filling the whole of the lower part of the picture, but also high, stretching from bottom to top of the picture; the farthest lintel, parallel to the pictorial plane, of the same compartment then becomes very low. Thus the lintels and ceilings of other compartments in distance are entirely blocked, and what visually remains of these compartments is dominantly the floors as a relatively clean backdrop for the two figures sitting on a rug, of whom the one in checkers sleeves is likely to be Shuten-dōji, one of the *oni* in Japanese folklore.[26] Masanobu even gently tilts up the floors of the distant compartments and tilts down the lintels of the most distant bay of the nearest compartment, in order, I suspect, to enhance the effect of the backdrop. Moving to the outdoor areas, on the right of the picture and flanked from top to bottom by the lintels and the outer edges of the deck of the nearest compartment, we find that the presentation of the landscape images appears to be "spatially" more consistent in itself, and is given a privileged position in the picture, almost in a mirror relation to Shuten-dōji. This approach is quite understandable: known as Mt. Ōe, this place of mountain and river is where Shuten-dōji's palatial abode is situated and thus very important in the mythical story; and precisely because of Shuten-dōji's occupation, no human activities are to be expected to take place here, hence the image of a landscape of mountains, rivers, cascades and trees.

Many other examples from Masanobu's oeuvre also demonstrate his tendency of taking flexible approaches to and making multiple choices for various parts of a single picture, which appear conspicuously at odds with the geometrical consistency of space in the post-Renaissance European pictorial tradition. In their remarkable studies of application and manipulation of linear perspective around the Edo period as a dialogue with Western modes of seeing, Yoriko Kobayashi-Sato and Mia M. Mochizuki insist that "from the start, the introduction of linear perspective in Japan was less about its correct application than about new ways of apprehending, conceptualizing and communicating the experience of 'world'."[27] Discussing the skewed positioning of the central screen in the "Chinese House" (*Tōjin-kan no zu*) as an examples of Masanobu's selective employment and manipulation of perspective, Kobayashi-Sato and Mochizuki state that "for Masanobu, the depiction of landscape was what was truly at stake," and therefore "he had no scruples about deviating from the laws of perspective." Here, as in many other cases, the artist's focus "was on integration, the blending of techniques to achieve a better solution," rather than on correct and accurate application of any single technique irrespective of cultural context.[28]

Masanobu was of course conversant with the rules of linear perspective and capable of producing images that follow its principal rules, as exemplified by the *Kabuki Theatre District in Sakai-chô and Fukiya-chô, a Large Perspective Picture* (Figure 3), probably made also in 1745. Thus where spatial inconsistency does occur, it must, at least in most of the cases, have resulted from Masanobu's intentional act for a certain desired effect that he wanted to achieve. The inconsistencies are drastic in some but subtle in some others (as shown in Figures 4–6, for instance), each nevertheless being specific in both treatment and purpose.

Approximately one century later, Katsushika Hokusai, best known for his *Thirty-six Views of Mt. Fuji* (*Fugaku sanjūrokkei*) created around 1830–32, in which *The Great Wave off Kanagawa* (*Kanagawa-oki nami ura*) has become an iconic print worldwide, was knowledgeable of and adept at linear perspective as much as, if not more than, Masanobu. And yet like Masanobu, Hokusai seems freely to employ linear perspective and to adjust it in ways that are against its European rules, whenever it deems appropriate to do so. Angus Lockyer, referring to the *The Famous Places on the Tokaido in One View* (*Tokaido meisho ichiran*), holds that "Hokusai saw no need to restrict himself to what he could see, much less to depict what he saw as it appeared before his eyes. ... Hokusai was at liberty to distort the map to make his point. Mt. Fuji not only occupied space in the physical landscape, it provided a spiritual anchor in relation to which the world was configured and a viewer could find their place."[29] I am particularly interested, however, in those examples in which contradictions and tensions in different modes of presentation are manifested. Take for instance the *Nihon-bashi in Edo* (*Edo Nihonbashi*) in Figure 7, from the series *Thirty-Six Views of Mt. Fuji*, in which frontal views, parallel-line presentation, and a mode that is similar to linear perspective are all present. Screech's reading of it is particularly insightful:

> The hubbub of the city is seen on the bridge in the foreground, rigidly confined and pinned into its proper subordinate scale; the lowly are arranged transversely so as to be unsusceptible to perspectival treatment. Beyond, the pompous warehouses of the city's merchant élite extend into the distance bearing their identifying markings; these are in perspective, which is not amiss for over-blowing is the merchants' nature ... At the rear of the print, though, the turrets of the shogunal castle and the peak of Fuji, the two great symbols of the realm, are discerned; not things of pride or vanity but the noble hubs of the Japanese state and of ancestral culture, these elements remain precisely not included in the perspective scheme: Castle and peak, creatures of an altogether grander dispensation, are shown as inaccessible by way of any of the parallels that unite what dwells beneath. The populace is crushed below, the Nation and its monuments spread out above; Western perspective is what governs the middle echelons.[30]

Screech's argument certainly holds, and I merely want to emphasize a point with regard to Hokusai's employment of perspective itself. Those "pompous warehouses," due to the intended direction of viewing, have to be rendered as stretching in depth, but should this situation be taken automatically as an issue of susceptibility to perspectival treatment? Depth can be expressed with other methods than linear perspective. In this instance, the depth along the canal flanked by warehouses forms a "vista" in the strict sense of the word, and these warehouses are treated with a suggestion of linear perspective; they are, however, far from being "in perspective," as shown in Figure 8. In this sense, the middle echelons are not really governed by Western perspective, but are taking advantage of it, thereby functioning as a stage on which this exploitable but alien mode of perception and comprehension is engaged.

On the other hand, the literal stage, instead of serving such a perceptual engagement, becomes a pictorial centre that is served by perspectival manipulation in another group of *ukiyo-e* prints which, often bearing the term *uki-e*, are also of interest to the present discussion. These are the pictures depicting scenes of entertainments of *kabuki* theatres. Kobayashi-Sato and Mochizuki cite an example from Masanobu's prints, *Large Perspective View of an Opening-of-the-Season Performance Onstage at the Kabuki Theatre* (*Shibai kyôgen butai kaomise ô-uki-e*), in which the orthogonal lines of the ceiling roughly converge on a single vanishing point, whereas other recession lines do not; hence the existence of what is called a "vanishing area," so as for the stage to be presented relatively larger. This, Kobayashi-Sato and Mochizuki reason, would appeal to prospective consumers, who, looking at the print, would feel as if they were physically in the theatre appreciating their favourite actors closely.[31] Kobayashi-Sato and Mochizuki then move on to a work by Utagawa Toyoharu (ca. 1735–1814), suggesting that the vanishing area has become more tidy because of clear grouping of different vanishing points.[32] Indeed, the majority of Toyoharu theatre prints that I have consulted show a shared pattern: orthogonal lines on the right side roughly converge at a point slightly right of the centre, and those on the left side at a point slightly left of the centre (Figure 9), a pictorial measure that comes close to what is called horizontal "fishbone" or "vanishing-axis" perspective.

Horizontal vanishing-axis perspective, known as *kubomi-e*, was also applied in images of outdoor settings, with buildings and other architectonic structures providing principal perspectival lines. Kobayashi-Sato and Mochizuki have cited a very good example from one of the scenes of Hokusai's illustration of *The Treasury of Loyal Retainers* (*Chūshingura*).[33] Another example is shown here in Figure 10. Hokusai even codified this representational approach into a principle of tripartite division of the picture plane—"a method of dividing a composition into three parts," as he calls it with two illustrations contained in the *Hokusai Manga* (Figure 11).[34] Theoretical codification does not necessarily mean that the rules were closely followed in application. In any event, even codified method of vanishing-axis perspective still maintains its power to rip apart a supposedly unified space, and consequently the space is no longer space as geometrically conceived and understood.

* * *

All these happened during the Edo period (1603–1868) when Japan had largely closed itself off from foreign influence, allowing only Chinese and Dutch to trade with it at the port city of Nagasaki from the mid-seventeenth century onward. This means that much of the introduction of Western pictorial techniques arrived in Japan was "filtered through its Chinese visual translation."[35] How much impact Nian Xiyao's (1671–1738) *Studies of Vision* (*Shixue*), first published in 1729 and reprinted in 1735, had really had on Japanese artists remains debatable, but it is certain that in appropriating Western conventions China had for a certain pe–riod taken the lead. One of the Chinese artists who are believed to have done so was Xu Yang, a native of Suzhou active around the mid-eighteenth century. Xu Yang was appointed first-class painter in the court academy after he presented his work in 1751 to the Qianlong Emperor who, on his first southern inspection tour, happened to be visiting Suzhou.

Xu Yang was particularly skilful at depicting extensive urban scenes, in both hanging-scroll and hand-scroll format. Let us first look at an example of hanging-scroll, *Poetic Ideas of Springtime in the Capital* (*Jingshi shengchun shiyi tu*), painted in 1767. "At first glance," as Anita Chung quite carefully states, "it seems that linear perspective has been adopted in the ordering of the pictorial space" of the painting, but it is, in fact, not really the case—each quarter of the palace, as well as each section of the urban areas in front of it, "is organized individually by means of parallel, converging, or diverging lines."[36] This, however, is only one aspect of Xu Yang's handling of the image, as there are many other notable "distortions" in the painting, most of them being clearly intentional. For instance, the buildings on the central axis, as well as those of the Imperial Ancestral Temple, are enlarged, an approach very much in line with the one in figure painting, where the central figure, highest in the social hierarchy, is usually rendered different in scale from other fig-ures in the same picture. The implied distance from the Gate of Heavenly Peace (Tiananmen) to the Meridian Gate (Wumen) is enlarged as compared to the distance from the Central South Gate (Zhengyangmen) to the Gate of Heavenly Peace, perhaps in order not only to present both the Gate of Heavenly Peace and the Meridian Gate more fully, but also to expose a decent portion of the area in front of each of these two prominent gates. Also enlarged, this time more significantly, is the distance from the Meridian Gate to the Gate of Supreme Harmony (Taihemen) and, still more significantly, from the Gate of Supreme Harmony to the Hall of Supreme Harmony (Taihedian), so that the most important court in the Forbidden City becomes highlighted. Xu Yang further emphasizes this court by slightly tilting it up, resulting in a perspectival "deviation" of their orthogonal lines from those of other parts of the Forbidden City. All these "deviations" and "distortions," along with other uncited examples, make futile a reading of the painting as presenting a unified and geometrically consistent space.

This kind of "deviations" and "distortions" are much more salient in Xu Yang's handscroll paintings, a format that renders the making of a unified and geometrically consistent space for the whole painting practically impossible.[37] But there is much more to it. Take for example the 1759 *Burgeoning Life in the Prosperous Age* (*Shengshi zisheng tu*), in which it is utterly meaningless to think of any vanishing point. In general, parallel-line method is applied to the presentation of buildings and architectonic structures in the entire painting, with directions of the front of buildings alternating between to the right and to the left depending either on their positions or the need of pictorial composition. As shown in Figure 12, where images of great depth—especially those of a street running to the picture plane at an acute angle—occur, the use of largely parallel lines still dominates, and the gradual diminution in scale of buildings in commensuration with the diminution of human figures is principally achieved by successively reducing the size of each of the individual buildings that form a row and extend to distance. This technique is in fact already present in Zhang Zeduan's *Along the River at the Qing-ming Festival* (*Qingming shang he tu*) painted in the early twelfth century, (Figure 13) where diminution in scale is effected by parallel lines and stepped reduction of height as buildings stretch from near to far. The only exceptions in the *Burgeoning Life in the Prosperous Age* are those images of a continuous structure, such as battlements on the city walls, outer walls of a granary or a long veranda, in which converging lines are used. In both cases, if we were to consider the diminution in size of a specific street scene stretching to distance as suggesting a particular "vanishing region," the coexistence of two or more different "vanishing regions" within a section—1 to 1.2 metres in length for viewing—of the scroll would pose a serious problem of spatial perception. It is precisely with this kind of spatial problem that Giuseppe Castiglione (1688–1766) had to come to terms by taking a perspectival system with three vanishing points in his 1728 *One Hundred Horses* (*Bai jun tu*), an approach that was, according to Marco Musillo, derived from "the early modern tradition of European theatrical imagery" (Figure 14).[38] But for Xu Yang, this should, we may surmise, not be problematic at all.

The same approach seems more systematically applied by Xu Yang a decade later to *The Qianlong Emperor's Southern Inspection Tour* (*Qianlong nanxun tu*), completed in 1770. Maxwell K. Hearn is right in indicating "the modified use of Western-style linear perspective" in this scroll by comparing it with Wang Hui's *The Kangxi Emperor's Southern Inspection Tour*, datable to 1698, but I am not sure if his insistence on "Xu Yang's adherence to Western conventions of spatial realism" and the artist's "commitment to maintaining a consistent point of view" can be fully valid, unless Hearn's use of "spatial realism" and "viewpoint" is only figurative.[39] We may take the section depicting the Changmen Street in Scroll Six, "Entering Suzhou along the Grand Canal" (Figure 15) to illustrate my point. The street is presented here as stretching in depth from bottom-left diagonally to middle-right, with clear diminution in scale along its length as if the roof ridges, eaves, balconies and baselines of the buildings flanking the street have constituted imaginarily continuous lines that appear to be converging. However, as Figure 16 shows, each of these seemingly continuous lines is in fact formed by a series of short, separate lines, and each of these short lines delineates the roof ridge, eave, balcony or baseline of a single building. Allowing for one or two odd lines suggestive of convergence, the lines for each individual building are basically parallel to each other. Xu Yang is clearly taking advantage of, or visually creating, the partition of the row of buildings on either side of the street, readjusting the position or direction of each line to help effect diminution in scale; hence a sense of convergence of lines to the modern eye. Thus, linear perspective is not really applied, and there is no vantage "point" for viewing the "space" of his urban scenes. Xu Yang certainly appropriates the European representation techniques, but my argument is that he might not have had any qualm about "deviating" from those techniques simply because he, like Masanobu and Hokusai, was not mentally or intellectually constrained by the conception of the world as a unified and geometrically consistent space.

* * *

The reason for incessantly fussing in modern scholarship over the issue of linear perspective, its accuracy, or lack of it in Chinese landscape painting, lies precisely in the taken-for-granted idea of space. It is in this line of thinking that I consider the constructed interpretative contrast between European and Chinese paintings, outlined earlier above, not just as a construct that is over-simplistic, but, at least potentially, one that is misleading. Even the word "projection" and the concept it signifies, such as the phrases "parallel projection" and "axonometric projection," becomes dubious in an interpretation of pre-modern Chinese landscape paintings. Needham holds that the Chinese "had been conscious of the problem of projection, how to represent three-dimensional space upon a plane surface."[40] Projection, however, should not be equated with two-dimensional representation of three-dimensional space, and no explicit qualification has been given by Needham for the use of the concept "projection" in the Chinese context.[41] The shadow of a gnomon would be cast on the ground and those of bamboos on the garden wall, but recognition and making use of these is conceptually different from application of what we now understand as "projection" on the theoretical and technical basis of descriptive geometry. Our approach to and appreciation of Chinese painting has indeed been taking firm place in a frame of reference that is alien to its tradition.

Indeed, there is no word in classical Chinese that denotes "space."[42] Its modern equivalent, *kongjian*, is borrowed, or re-adopted, from modern Japanese in early twentieth century. The Japanese words for both "space" (*kūkan*) and "time" (*jikan*), formed with Chinese ideographs, were coined in the Meiji period (1868–1912) to correspond to the new concepts imported from the West, and, as suggested by Dōshin Satō, in Japan "the awareness or concept of space was not linguistically coded until the emergence of the word *kūkan*."[43] In classical Chinese and later in Japanese, *kong* is usually taken as referring to void or emptiness, but *jian*, originally composed of the ideographs for *men* (door) and *yue* (moon), is "defined" by Xu Shen (ca. 30–ca. 124) as cleft or interstice, and Xu Kai (920–974) comments on it in a figurative way: "The door at night is closed and yet the moonlight appears [inside the room], then there is an interstice."[44] This ideograph has another, different but related, meaning: "unoccupied" and, derivatively, "slightly in leisure."[45] Thus *jian* basically means "in-between," either of things or of conditions.

The modern concept of space covers as much the world of our everyday life as the "universe" and "cosmos," and the pre-modern Chinese concept that comes close to the latter is expressed with the words *tiandi* (heaven and earth) and *yuzhou*. *Tiandi* seems less controversial in translation than *yuzhou*. There are two sentences in the "Gengsang chu" section of the *Zhuangzi*, which explains *dao* with *yu* and *zhou*. Burton Watson's translation of them is as follows

> It has reality yet there is no place where it resides —this refers to the dimension of space. It has duration but no beginning or end—this refers to the dimension of time.[46]

I cannot help but wonder whether the imposition of modern concepts "dimension of space" and "dimension of time" on *yu* and *zhou* respectively may not help us as much understand as misunderstand the original statement. But the question remains as to "dimension" of what if not "space" and "time," for which I admit that I have as yet no answer. Two different but resonant sentences, contained in the chapter entitled "Qisu xun" of the *Huainanzi*, are translated by Andrew Meyer in the following way:

> From furthest antiquity to the present days is called "extension-in-time";
> The four directions [plus] up and down are called "extension-in-space."[47]

This translation brings to me a similar discomfort, although it attempts to facilitate modern understanding of a text written nearly two thousand years ago. The two ideographs are highly figurative in these cited contexts and thus, as in the cases of most concepts in Chinese literary tradition, far removed from giving any sense of abstract definition. In the chapter "Lanming xun" of the *Huainanzi*, the same ideographs are used in the combined form and this time each refers to its original meaning—the edges and ridgepole of a building.[48] Imposing abstract concepts, which is familiar to the modern mind, on figurative statements unfortunately has the tendency of distorting the original authors' conception and perception of the world. In any event, my point is that what we understand now as "space" is absent in the pre-modern Chinese textual tradition, in either experiential or cosmic dimensions, and what we confront in this tradition is a world without "space," in which landscape and garden are situated.

Moving from the largest scale of the universe to one of our everyday life, one realises more clearly how deeply entrenched the concept of space has been in modern mind. Habitual or *a priori* application of "space" in translations abound, a brief note of a few examples from highly respected scholars should suffice. Richard E. Strassberg, for instance, solicitously chooses the wording for one of the famous lines in the text when translating Ouyang Xiu's (1007–1072) "Record of the Pavilion of the Old Drunkard" (Zuiwengting ji, 1046): "But wine is not uppermost in the Old Drunkard's mind. What he cares about is to be amid mountains and streams" 醉翁之意不在酒 在乎山水之間也. And yet remarkably sensitive to cultural differences as he is, Strassberg has to render the words 兩崖相廞間 as "in the *space* between the two cliffs" in his translation of Bai Juyi's (772–846) "Preface to Poems from the Cave of the Three Travelers" (Sanyoudong xu, 819). Missing the two characters 相廞 in ranslation aside, even though the word "space" may not be absolutely necessary here, it nevertheless causes no serious problem, as it means "extent," "interval" or "distance." But it does when it appears in a passage in Liu Zongyuan's (773–819) "The Little Hill West of Flatiron Pond" (Gumutan xiaoqiu ji, 809):

> I lay down using a mat as a pillow. Clear and cool forms sought out my eyes, the gurgling sound of water sought out my ears, *the expansive*

> *space* sought out my spirit, and the capacious quietude sought out my mind.[49]
> 枕蓆而臥 則清泠之狀與目謀 瀯瀯之聲與耳謀 悠然而虛者與神謀 淵然而靜者與心謀

In a modern mode of perception, that “the expansive space” communicates with one’s spirit certainly makes sense, but was it really that space that Liu Zongyuan felt to have communicated with his spirit?

In Chinese art history, the employment of “space” similarly happens in great frequency. James Cahill, for example, translates the words in Zheng Sixiao’s (1241–1318) casual seal on his *Orchid* (*Molan tujuan*, 1306) as follows:

> If you ask me, you won’t get it; if you don’t ask, I may give it to you. In the *great space and expanse* before my eyes, there’s only the pure breeze of the present and the past.
> 求則不得 不求或與 老眼空闊 清風萬古

As Cahill has emphasized, this is“freely rendered”; indeed, especially in Cahill’s construal of the first two lines of the original text, which seem to reveal Zheng Sixiao’s understanding of the principle of life and nature, rather than possible situations of engagement between the “I” and “you,” and in missing out the old age of Zheng’s eyes, which is crucial for bringing out the force of the last four characters. But judging from Cahill’s otherwise elucidating discussions throughout the whole volume, the use of the word “space” here, clearly inaccurate and misleading, is most unlikely to be casual.[50]

The problem exists as much in freely rendered texts as in those of rigorous translation. Let me cite one example—the readings of a well-known sentence from Guo Ruoxu’s (2nd half of 11th century) *Tuhua jianwen zhi* (ca. 1080): *shenyuan toukong, yiqu baixie* 深遠透空 一去百斜.[51] Alexander Coburn Soper’s translation runs,

> Perspective distances will penetrate the space, with a hundred [lines] converging on a single point.[52]

Linear perspective is taken as a guiding principle for reading the words of scholar in eleventh-century China, suggesting that the post-Renaissance European pictorial convention be universally applicable. The translation by Susan Bush and Hsio-yen Shih appears to distance itself from linear perspective:

> Deep distances penetrate into space and a hundred diagonals recede from a single point.[52]

The paradox is that inherent contradictions are created in this rendering by Bush and Shih precisely because any overt reference to linear perspective is removed and yet the words "space" and "a single point" are nevertheless retained in such a way as to be evocative of linear perspective, contributing to the fact that the second half of the sentence now makes little sense.

Robert J. Maeda thoughtfully combines this sentence with the one immediately precedes it and keeps his translation clear of any overtone of linear perspective:

> ... and brushstrokes of even strength should deeply penetrate space, receding in a hundred diagonal lines.[53]

Maeda's interpretation of the sentence is particular insightful. In any event, with or without an implication of linear perspective, the character *kong* (void, emptiness) has been turned into "space" in the above cited examples. But we know that *kong* does not mean "space," and I suspect that these prominent scholars do as well, at least when this particular character were to be the focus of their studies. It is precisely in this line of thinking that I find Caswell's translation judiciously free of an either perspectival or spatial burden:

> Deep distance penetrates the *voids* and together yields many slanting or dispersed [views].[55]

Scholars' writings on gardens display a similar tendency in translation. Duncan M. Campbell's translation of a passage in Li Dou's (1749–1817) *Yangzhou huafang lu* reads: "Touring this *space* one feels oneself to be like an ant crawling through the twisting eye of a pearl, ... "[56] 遊其間者如蟻穿九曲 . Campbell has always been noted for his admirable skills in smooth translation, but in this case, as briefly explained earlier above, *jian* in pre-modern Chinese literature does not mean, and not even imply, "space." One last example comes from Wai-Yee Li's translation of the section on the "Turning Bracelet" (Wanzhuan huan) in Qi Biaojia's (1602–1645) "Notes on Allegory Mountain" (Yushan zhu) likewise contains such a sentence: "Here one starts walking and in no time reaches the summit of the Distant Lodge (Yuange): this is Old Man Gourd's (Hugong) magical art of contracting *space*." 兹纔一舉步 趾已及遠閣之巔 是壺公之縮地也.[57] Here, rendering *di* 地, which should more adequately read as "place," as "space" to some extent seems not only to distort the original meaning, but also to deprive it of possible social and experiential overtones.

These examples, all taken from the works of respected scholars, are certainly not meant for any disservice to their admirable achievements. Originally trained in architecture, I myself have firmly remained in the orbit of modern spatial concept, and my previous writings are replete with the word "space"; but using mine would be much less persuasive than citing those of these scholars'. Thus in so doing I here am merely demonstrating how deep-rooted and pervasive the concept of "space" is in our perception and thinking of Chinese painting, garden and landscape. Even the schemata of the famous "three distances" (*sanyuan*) formulated by Guo Xi (ca. 1020–ca. 1090) in his *Lofty Ambition in Forests and Streams* (*Linquan gaozhi*) have sometimes been enlisted as ways of spatial perception. The fact, however, is that they explicitly concern mountains only. For this reason, one may have to be alarmed when the schemata are even just moderately expanded by Wen Fong, a prominent art historian of tremendous accuracy and sensitivity, to be "ways of perceiving landscape."[58]

One may still argue, as Mitrović possibly would when he methodologically warns us not to "confuse the concepts used in the analysis with the assumptions these concepts are meant to analyse," and also insists that "the absence of a theoretical concept cannot be taken for the unawareness of the phenomena the concept was subsequently constructed to explain."[59] But the concept of "space" has not merely been used in the analyses; it often frames and shapes our perception. It can also hardly be the case that Li Cheng and his contemporary painters in the tenth century had already had an awareness of the phenomenon of "space," and yet it took about one thousand years of continuous development in painting for the concept suddenly to emerge around the turn of the twentieth century to explain this phenomenon, while the specific term for this concept had to be borrowed from modern Japanese, in which this term was coined only in the late nineteenth century.

* * *

Terms in classical Chinese, apart from those we have already encountered above, which can be related to our modern concept of space are numerous. Some of them can find in English their equivalents, either close or distant, while others cannot. Most of these terms signify or imply position, orientation, situation, and interrelation; and a great proportion of them are in conceptual polarity (*xiang/bei, gao/xia, yuan/jin, xia/kuo, xu/shi*, etc.), whether the other of the pair is present or absent in a specific context. Even a word like *guangmao*, often translated into one all-encompassing English word or phrase, such as "expanse" or "great expanse," is in fact polarized, as *guang* refers to the east-west direction and distance, while *mao* the north-south. As the longest living civilization on earth, China has been vital rather than moribund in history, which necessitates continuity as much as change, in a way characterized nicely by Pierre Ryckmans in allusion to Parmenides and Heraclitus: "permanence does not negate change, it informs change."[60] Such continuity also manifests in the mode of perception of the world, which has developed from the Han to the late Qing periods.

Evidence can never be mustered sufficiently to demonstrate a general lack of a phenomenon in a culture, and citing examples of "what is not" merely functions to illustrate an argument. For instance, the well-known passage on the canonical city plan in the *Records on Investigating Crafts* (*Kaogong ji*) in the *Rites of Zhou* (*Zhou li*) reads:

> The artificers, as they built the capital, set it as a square with sides of nine *li*, each side having three gateways. Within the capital there were nine meridional and nine latitudinal avenues,

> while each meridional avenue had nine chariot-tracks. On the left is the Ancestral Hall and on the right the Altar to the God of the Earth; in the front is the Imperial Court, and at the back the Market.[61]

What impinges on us of the quoted passage are configuration, number, orientation and positional interrelation, without any implication of space.

This mapping mode of description, emphasizing orientation and positional interrelation, seems to have persisted so unwaveringly that it continues to work in both travel and garden essays from the Northern Song period (960–1127) onward, albeit in more elaborative ways on the one hand and combined with a new descriptive mode, the mode of touring, on the other.[62] Zhu Changwen's (1039–1098) "Record of the Pleasure Patch" (Lepu ji) is, in this respect, a pertinent example of the garden essays.[63] The layout of this garden is presented in a designed combination of mapping and touring modes, directing the reader's attention from one domain to another: the residential quarter dominated by a hall of three bays is briefly mentioned first; this leads to a distinctive area south of the residential quarter, characterized by the Hall of Limitless Classics (Suijingtang), also of three bays; the reader is then introduced, from the northwest corner of the Hall of Limitless Classics, to the Ridge for Seeing the Mountains (Jianshangang) and the pool at the foot of the hill; to the north of this hill lies the West Patch (Xipu); and finally to the southwest of the West Patch arises another, natural hillock called West Mound (Xiqiu). Thus the presentation of the garden proceeds through a process of successive unfolding of the five domains with regard to their directional and positional interrelations, while each domain, except the last one, is in turn centred either on a significant artificial structure or a natural feature functioning as a reference point for delineating the directional and positional interrelations between structures within the specific domain, allowing occasional cross-referencing between two domains only for the purpose of ideational input rather than for spatial definition. Unsurprisingly, none of these directional and positional interrelations is quantified.

The primacy of directional and positional interrelation in

the Chinese mode of perception is particularly evident in picture maps. Take for example the 1229 *Picture-map of Pingjiang* (*Pingjiang tu*), which, like the vast majority of city maps in pre-modern China, employs a representational technique that seems to "explode" each building compound and any other socially or culturally recognized area. But it is fundamentally different from what we know as "exploded view drawings." The latter is generated by following geometric rules and presenting an object logically dissected in a unified space and viewed from a certain fixed vantage point. In contrast, the *Picture-map of Pingjiang* presents multiple series of unfolding domains. As exemplified by the central walled enceinte (*zicheng*) (Figure 17), the frontal images of buildings in each domain are seen as lying supine in front and on the right and left of the imagined visitor moving from one domain to another, with some of the double-leaf doors at large gateways—the main gate in the south wall and the side gate in the west, for example—being kept ajar, while the roof surfaces of the corridors connecting buildings along the central axis is shown as if seen from above. The representation of the Justice Office (Tixingsi) and surrounding streets in Figure 18 is even more revealing—entrance to the office compound is on the east end, and the frontal images of the buildings bordering the first court on its north, south and west sides are lying supine as if each being seen by the visitor in the court; but coming to the second court, the visitor finds the building on the south side lying supine but upside down relative to the visitor's position, which means that this particular building may not be part of the compound in terms of use and its front and entrance are facing south to the street outside of the compound. Still another example from the *Picture-map of Pingjiang* comes from two small compounds, the Juyang Almshouse (Juyangyuan) and the 41st Weiguo Barracks (Weiguo sishiyiying), to the west of the Southern Star Bridge (Nanxingqiao) at the foot of the south wall of the city (Figure 19). Having their entrances in the north, the two compounds are different in the orientation of buildings inside: the presented layout of the almshouse is straightforward as most compounds in the picture-map, but in the image of the barracks, the buildings inside the compound turn their front southward, which is readily understandable—they are buildings to house soldiers and for biological and cultural reasons they should face south, whereas the buildings in the almshouse did not function as routine human residence and thus did not need such a conventional orientation.

The vision of the invisible visitor, who is supposed to follow the proper or ordinary routes, remains largely within the picture-map and the directional and positional interrelations of the buildings are presented by reference to the visitor's moving positions along the routes, except for the cases of the roof surfaces, in which the image refers to the position of the viewer outside of the picture-map looking at it in distance; in other words, what the viewer sees in the picture-map is, to a great extent, the vision of the invisible visitor in the picture-map. In garden painting and drawings, however, the visible visitor becomes part of the view; what the viewer sees in the painting or drawing is different from the vision of the visitor in the image. The wood-block-printed handscroll *Illustrations of the Hall Surrounded by Jade* (*Huancuitang yuanjing tu*) drawn by Qian Gong (fl. 1579–1616) offers us a good example. Figure 20 contains the sections that depict four adjacent domains, namely the front court of the "Hall Surrounded by Jade," the "Remaining Splendour of the Orchid Pavilion" (Lanting yisheng), the "Spring Gushing to Heaven" (Chongtianquan), and the "Terrace of Achieved Life" (Dashengtai) with the "Tower of a Hundred Cranes" (Baihelou). Here even untrained eyes should notice the "spatial" incongruity between all these four domains. They are congruous with each other, however, only in terms of their relative positions and mutual connections through the walls, steps and entrances, of the scales of buildings and human figures, and of the placement of trees and vegetation.[64] Within each visually presented domain, no particular vantage point can be determined; or put it another way, the idea of vantage point and linear perspective is entirely irrelevant to this mode of perception, and so is the concept of parallel "projection." In short, the concept of an infinite, isotropic and homogeneous space has no place either in or for it.

I am not denying the instrumental value of the modern concept of space in analysing Chinese garden and landscape, which has proved particularly useful to our thinking of modern architectural and garden design. Nor am I even proposing that the word "space" be shunned in discussing pre-modern China art, architecture, garden, and so forth. I merely suggest that if we are trying to understand how pre-modern Chinese garden and landscape painting were conceived by their creators and perceived by their viewers, it should be important to remember that this modern concept was absent in their mind. How one perceives is basically coterminous with how one thinks, and as Massimo Scolari has emphasized when discussing the Jesuit's attempts to introduce perspective to China, "changing the way of seeing, and therefore of representing, meant changing the mode of thinking."[65] By imposing our modern concept on the pre-modern Chinese conception and perception, we in effect alter their mode of thinking; hence the distortion. This is not a question of the Chinese either pictorially or textually "*mentioning* the subject of space," with which John Hay opens his remarkable discussions of "Chinese Space in Chinese Painting"; it is, instead, a question of the very existence of the concept and of particularity of the mode of thinking. But I take Hay's thoughtful qualification most seriously: for Hay's convenience of discussion, "space is simply extension, that which is necessarily implied by any kind of organization. In other words, that it is the medium of existence."[66] The pre-modern Chinese kind of organization is surely different from the European one; in the Chinese context, such "extension" does not necessarily extend into the homogenously infinite, while "the medium of existence" does not have to mediate with measurements and unification. The hills and ravines in the heart of those painters and garden designers constitute a world of its own, distinctively free of geometrical space.

胸中丘壑：一方没有“空间”的天地

许亦农

（韩阳 译）

刘敦桢（1897—1968年）在1957年发表的《苏州的园林》一文中，用“空间”这个在现代汉语中与space相对应的术语来讨论园林中的“空间组合”问题。[1] 尽管该词在这篇文章中的使用有限，但它仍然展示了一种特定的话语取向。这种取向在刘敦桢逝后1979年出版的、颇具影响力的《苏州古典园林》一书中得以强化，不仅在“总论”的第二章“布局”中包含专门讨论空间概念的“景区和空间”一节，而且这个概念从根本上贯穿全书。[2] 这一理论的发展势头就此开始，“空间”成为其后中国建筑史家在园林研究中的一个核心的、想当然的理念，而有些研究思路甚至把它当作一个参照系，致使园林讨论完全在这个理念的统摄下进行。

刘敦桢著述的发展轨迹与童寯（1900—1983年）的形成鲜明对比。童寯当然绝非不熟悉“空间”的概念，[3] 但“空间”这个词在他发表于1936年的文章《中国园林：以江苏和浙江两省园林为主》中仅用过一次，在很大程度上意味着“区域”或“场地”；[4] 而其1937年完成的经典作品《江南园林志》虽然颇具传统主义意味，展示的却是一种对江南园林的现代解读，尽管如此，“空间”或隐含这个概念的任何词语都未出现。[5] 童寯在1978年还写了一篇题为《苏州园林》的英文文章，目的是为刘敦桢的《苏州古典园林》作序；有趣的是，他在这篇文章中开始写下这样的句子：在园林中“曲线和考究的不规则……乃其设计特点，空间布局将视野限于某个单一如画庭院”以及“悉心地通过开敞与封闭空间……而达到对比的效果。”[6]

刘敦桢和童寯是公认的中国现代建筑史学的先驱。这里简短提及两位亲爱的老师的著作，目的并非比较他们个人的园林研究方法，[7] 也不是暗示其研究思路代表了他们那个时代出现的所有园林方面的研究，而是意在设定一个起点，以探讨可能存在的对中国园林的误解。这个起点隐含着几点认识，即国内讨论园林设计时使用“空间”概念作为参照系是一种现代现象，这种现象始于建筑学的现代发展，[8] 而“空间”或许不是唯一的感知方式，甚至不是一个主要方式。中国园林一直具有强烈的“建筑性”，大多数园林方面的学术研究都在建筑院校中进行。因此诸如“空间”之类的现代建筑概念和术语自然被引入园林解读之中。

然而，事情似乎并不这么简单。造园与读园从本质上是人文学科的一部分，从而也是文化理论的一部分。既然如此，借用克利福德·格尔兹（Clifford Geertz）的话来说，园林方面的分析和讨论“不是寻求规律的实验性科学，而是寻求意义的诠释性科学。”[9] 其语言和词汇也因此必须或多或少地分享人文学科不同领域中的语言和词汇。这一点对于现代到来之前的中国来讲尤其如此；正如牟复礼（F.W. Mote）所说，在这个“一以贯之的文明”中，“许多关键术语以明显的平行方式使用，并且一方面的活动中存在的价值很容易转换为其他方面活动的价值。”[10] 因此，人们在诗文、书画和哲学研究中会提出的问题、会使用的术语，同样应用于园林研究。造园和山水绘画之间的联系尤其紧密：两个明确独立的学科息息相关，有时甚至可以相互转换，彼此频繁地影响并丰富着对方，这种相互关联性和相互转换性使平行地解读山水画和园林成为可能。然而，正如下文中几个选例所示，空间概念之应用于现代到来之前的山水画的研究，很难归结为现代建筑的影响，而是来自一个更为普遍的现代的世界观。

空间已成为一个包罗万象的概念，那么在这篇文章中我想提出的问题是，就我们对于园林和山水画的理解而言，这个概念的现代注入是否既有利又有害。也就是说，尽管其本身的价值不可估量，空间具有排除其他解读可能性的倾向，从而给更好地理解带来障碍。近年来出现的挑战现代习惯性解读中国园林和山水画方面的学术尝试令人振奋，[11] 但这些挑战的弱点恰恰在于它们完全在同一概念框架内运作，其中所有的概念——包括这些挑战的目标——都是相互关联的，因此产生某种自我否定的效果，从而限制了其论证的力度。

* * *

在典型的中国山水画中，“画面空间”（pictorial space）既是现代术语[12]，也是现代概念，对其构建的广泛解释是观看发生在具有众多或散在或转移视点（multiple, dispersed, shifting viewpoints）的“透视”中，也可简称为多点或散点或移点透视（multiple, dispersed, shifting perspectives）；近大远小明显存在，以及后退物体的对比度、细部处理和色彩饱和度等渐弱的空气透视（atmospheric perspective）；在描绘家具、建筑物和其他构筑物时，通常会遵循平行投影（parallel projections）的原则。这种以“视点”、“透视”和“投影”为中心的解释，无疑是以貌似对立面的方式、参照后文艺复兴时期欧洲绘画研究而提出来的。除了通过某种特定类型的平行投影呈现的技术图像之外，人们一般认为观赏欧洲绘画是以一个固定视点、通过控制整个画面的线性透视为条件，从而创造了无限的、各向同性的、各处同质的空间，此空间则成为被描绘的物体所在的连贯一致的媒介；也就是欧文·潘诺夫斯基（Erwin Panofsky）所假设的“一种完全‘透视’的空间观。”[13] 在这里，空间连贯一致性在感知中至关重要。透视“作为画面空间的生成者”而运作，“适用于整个图画，而不仅仅适用于图画中的单个物体”；[14] 威廉·V·顿宁（William V. Dunning）言简意赅地指出，其最重要的功能是“创造一个统一的理性空间”。[15] 自从线性透视于十七世纪起被引入中国之后，两种传统之间的对比似乎变得更加醒目，因为这两种感知模式在其后的中国图像世界中经常被视为共存。

这种诠释性的对比确实具有吸引力且容易理解，但实际上是通过删减的方式而使欧洲绘画传统过于单纯化。[16] 由于这篇文章关注的并不是欧洲空间概念的复杂性、透视的发展或感知的历史，所以在此只需认识到，近三十年来关于艺术和建筑领域中透视发展问题的学术争论，在如下几个方面既提升了我们的意识，也增长了我们的知识：十五世纪意大利出现的“线性透视的重新发现”的历史构建；被看作是线性透视出现的一个必要条件的同质性空间构想；以及空间概念本身。[17] 不过与我们在这里的讨论更相关的是我们已经认识到，从十八世纪开始，欧洲绘画中关于空间的想法从未静止过；诺曼·布列逊（Norman Bryson）坚持认为欧洲绘画中关于空间的最普遍但错误的信念之一就是“从乔托（Giotto，约1267—1337年）到塞尚（Cézanne，1836—1906年），‘十五世纪意大利文艺复兴’空间一直无可争辩地占据统治地位”；他指出了许多打破阿尔伯蒂常规的历史案例，即使仅在法国，就包括安托万－让·格罗斯（Antoine-Jean Gros，1771—1835年）、让－奥古斯特－多米尼克·安格尔（Jean-Auguste-Dominique Ingres，1780—1867年）、佛迪南德·维克多·额尔金那·德拉克华（Ferdinand Victor Eugène Delacroix，1798—1863年）和洛可可绘画。[18]

不过我们不能否认，那极具影响力的各处同质、自成系统的空间概念在文艺复兴之后的欧洲的存在和应用，以及有助于创造这一概念的线性透视的存在和应用，尽管这只是绘画的众多方法之一；这也就为构建中、欧绘画的诠释性对比奠定了概念性基础，而这个基础继而又成为解读中国山水画的一个参照系。然而，更加仔细审视中国的图画经验时，我们不禁会怀疑这种诠释性对比或许不只是一个过度简化的思考方式。一些包括李约瑟（Joseph Needham）和詹姆斯·O·卡斯韦尔（James O. Caswell）在内的学者已经表示出这样的怀疑，尤其是针对透视这个概念的怀疑。李约瑟对中国和欧洲之间的知识、文化和历史差异有着令人钦佩的敏感度，他认为中国的制图（与绘画）的常规遵循他所谓的“扩散性视点区域的原则”（the diffuse view-region principle），在这个原则下“有一个‘多个位置的视点’，或者更确切地说是‘悬浮的或动态的观看区域’（hovering or dynamic view-region），而根本不是一个视点。”[19] 尽管李约瑟的观察依然是在欧洲绘画传统的基础上建立起来的参照系内运作，但却异常敏锐，以至于几乎要脱离这一系统。卡斯韦尔显然想当然地在他的陈述中采用同样的参照系：“……中国人对透视的某些运用并非一无所知，但是……他们为自己的目的开发了自己的方法”，可是他马上又默然而颇有见地地转向了与李约瑟的思路相似的论点：“对于中国绘画来说，‘透视’这个术语不甚合适，因为它承载着西方观念的负担。”[20]

李约瑟和卡斯韦尔的各自论点都在靠近问题的关键，引导我们去质疑那引发视点和透视问题的基本假设——山水绘画中呈现的图像必须从某个或某些视点来观看，这自然应该引发这样或那样的透视。但事实是，在细看诸如传为李成（919—约967年）所做的《晴峦萧寺图》的山水画中的图像时，“众多或散在或转移的视点”其实并不存在；也就是说，任何视点如要真正存在，无论其数量多么大，散布得多么广泛、移位多么频繁，其明确界定的位置都需精准地找出，但这幅画作的情况并非如此。因此，谈论“众多或散在或转移的视点”、甚至“悬浮的或动态的观看区域”仅仅是近似的说法，接近线性透视的范围，以便在既定的参照系中继续进行有逻辑性的讨论。从这个意义上说，虽然不想冒犯卡斯韦尔，但“透视”作为一个术语用于中国绘画时，则不仅只是不甚合适，而更确切地说是与中国绘画无关；它所承载的负担确实非常沉重，而且毫无根据。

从十八世纪以降，欧洲各种再现模式和绘画技巧的引入得以强化，这使中国的绘画实践发生了一些值得注意的变化。清代宫廷画家的一些作品和大量木刻版画都说明了这一点。

这些作品采用了线性透视，尽管在有些例子中得到充分或基本充分的运用，而在另一些例子里仅仅在局部上或表面上使用。[21] 正是那些局部性和表面性的例子为我提供了另一个机会来说明问题。类似的情况在浮世绘中也能看到，这种十七世纪到十九世纪蓬勃发展的日本艺术类型中的一些例子也可以被提及，以支持我的论点。征引浮世绘之例是合理的，因为许多相互联系的、早已建立起来的日本绘画传统，其中尤以大和绘和唐绘为主，都或多或少从不同历史时期的中国绘画传统中发展而来，并且一直都在日本艺术家的掌握之中，而自十八世纪三十年代起他们中的一些人已经开始采用线性透视。在此期间中国和日本的例子很多，但篇幅的限制只允许我引用寥寥几幅作品，首先是两位浮世绘艺术家——奥村政信（1686—1764 年）和葛饰北斋（1760—1849 年）的木版画，然后是清代艺术家徐扬（活跃期 1751—约 1776 年）的画卷。

* * *

奥村政信尽管以许多重要的技术发明或发展而闻名，例如红刷绘、柱绘和漆绘，但他在普及所谓的透视木版画——即浮绘——方面的重要作用也得到认可。[22] 奥村政信的著名作品之一题为《两国桥夕凉见 浮绘根元》，创作于 1745 年左右（图 1）。在这张画作中，建筑物的内部看起来是透视的，而外部场景则不然。缇门·斯克里奇（Timon Screech）认为“日本建筑……在某种程度上最适合用透视法去表达”，因为梁柱木框架结构提供了参照线，而编织的榻榻米垫为轻易绘制景象退远提供标记，然而对日本艺术家来说“相比之下，室外空间仍然更具挑战性。”[23] 与此思路相仿，斯蒂芬·利特尔（Stephen Little）的评论更加断然：

> 奥村政信利用他对单点和多点透视的理解创造了许多户外场景。这些尝试经常揭示出基本的误解，这是由于他不完全理解那些透视规则，也不甚明白在原本欧洲语境下支撑这些规则的本体论假设。[24]

从本体论上讲，奥村政信构想的图像的前提假设可能确实与欧洲语境中的非常不同，但我并不十分相信画作中所示的内部与外部之间的明显差异应该被解释为奥村政信无能或缺乏理解的表现。首先，此画面的内部透视仅仅是粗略地暗示出来，因为只有三组垂直于画面的线条汇聚在单一灭点上，而其他线则不然。其次，在其大概同一年创作的《风雅女酒吞童子 浮绘根元》（图2）中，内部和外部之间的“空间”差异同样显著，但形式完全相反——就欧洲透视常规的标准而言，室外景致以“空间上”相当合乎逻辑的方式呈现，而建筑内部虽然明显令人联想到透视的使用，却处于混乱状态。因此，认为奥村政信是否有能力将线性透视正确地应用于图像（无论是室内还是室外），实在说不通。

另一种解释可以是，奥村政信（或其他浮世绘艺术家）在作画时有宽广的选择范围。他不仅为单张图片的不同部分选择性地采用了不同的传统手法，而且还自由地调整任何单个传统手法的规则，以使图像最充分地传达预期的信息。就第一张图片而言，建筑内部有两个主要的活动场景，一个处在前景的中心，另一个位于右侧相邻的居室。透视方面的不一致可以这样理解：由分隔中央和右侧居室的过梁形成的线条以不同的角度向上倾斜，以使右侧居室的天花尽可能显露出来，但是由右侧居室右边缘的过梁形成的线条向下倾斜，在某种程度上有助于封闭该区域，从而使远处的河流图像转变为背景。在建筑外部，即使画面左边的木桥、桥下小船、画面中心位置的篷船构成了中景，但整个户外区域都被当作一个整体而处理，因此从活动和图像两方面讲都起着背景的作用。河流、桥梁和开阔的场地向上倾斜，目的很可能是为了清楚地显示场景中的所有户外活动——划船、过桥、餐馆打尖等等——这与室内悠闲放松的气氛形成对比；后者是个自成一体的世界，与外面繁忙的日常生活隔离开来。这种解释似乎恰好与斯克里齐的敏锐观察产生共鸣：

> 利用透视而显示的恰恰不是那些日常的、
> 被赐予的事物，而是舞台上的模仿空间、

> 消遣或逃避现实世界的“他者”、或现实以外的娱乐区域。
>
> ……浮绘与自负和浮夸串通在一起。它们可以被纳入叙事语境中；很多有插图的故事与传统空间构建方式的使用自在地并行，只是在引出一个浮华或张狂的空间时才切换到（常常即时采用）透视画法。[25]

所以浮世绘中透视的本体论功能，正如斯克里奇所论证的那样，与其在欧洲绘画中的功能正好相反。对透视功能的这种逆转的理解为我的论点提供了支持，即线性透视被当作一种新的、外来的技巧，任意取来而与传统的“宋元”和“大和绘”等图像表现形式并存。

然而在《风雅女酒呑童子 浮绘根元》中，构成前景的室内活动仅集中在距画面最近的居室里。为了提供足够的空间和良好的场面调度，此三开间居室外加纵深方向更远一开间区域的过梁和天花在“透视”上与远处居室的过梁和天花截然不同。靠近观看者的所有人物聚集的前景区域不但宽阔，填满了图片的整个下部，而且高大，从图片的底端一直伸展到顶端；那么这个主体区域最远处、与画面平行的过梁就变得非常低。远处其他居室所有的过梁和天花因而被完全遮住，使得这些居室在视觉上只剩下地板，为坐在地毯上的两个主要人物提供了相对干净的背景，而其中身穿格子袖笼者可能是酒呑童子——日本民间传说中的诸“鬼”之一。[26] 奥村政信甚至使远处居室的地板微微向上倾斜，最近居室中最远的过梁则稍稍向下倾斜，我猜想这是为了增强背景效果。转而再看图片右侧的室外区域，从顶部到底部由距画面最近的居室的过梁和露台外部边缘限定，这里我们注意到，其山水风景自成一体，在“空间”上似乎呈现得更加一致，在图片中享有颇为重要的位置，几乎与酒呑童子呈对影关系。该处理方法是完全可以理解的：这个山川之地被称为大江山，是酒呑童子的宫殿处所，所以在神话故事中非常重要；正是由于酒吞童子在此盘踞，这里不会发生任何人类活动，因此我们看到的仅是山峦、河流、瀑布和林木的景象。

奥村政信作品中很多其他例子也表明他倾向于用多种选择方式灵活处理单张图片的各个部分，这与文艺复兴后期欧洲绘画传统中空间的几何一致性明显相悖。小林頼子和米尔・M・望月みや对江户时代线性透视的应用和熟练操作进行的研究引人注目，其研究旨在与西方观看模式进行对话；她们坚持认为“从一开始，线性透视引入日本，并非为了正确运用它，而是在于认识、理解并传达人们体验‘世界’的一些新的方法。”[27] 小林頼子和望月みや将《唐人馆之图》里中央屏风的偏斜定位看作奥村政信选择性地使用并操纵透视的一个例子，并指出：“对奥村政信来说，描绘风景才是真正的关键，”因此，“他对偏离透视规则毫无顾忌。”与其他许多情况一样，这里艺术家所关注的是“专注于整合——通过不同画法的融合来实现更好的解决方案”，而不是正确、精准地应用任何单一画法，无论其文化背景如何。[28]

奥村政信当然精通线性透视规则，并且能够遵循其主要规则创作图像，例如很可能也是作于 1745 年的《境町葺屋町芝居町 大浮绘》（图 3）。因此，当空间上不一致的情况的确出现时，至少在大多数情况下，一定是奥村政信有意为之，以便达到其想要实现的某种效果。奥村政信作品中某些空间上的不一致颇为强烈，而另一些则相当微妙（如图 4-图 6 所示），无论哪种情况都有其特定的处理方式和目的。

大约一个世纪之后，葛饰北斋以其《富岳三十六景》而著称，其中《神奈川冲浪里》已经成为全世界标志性的印刷品。他至少与奥村政信一样熟知并擅长线性透视。然而，与奥村政信一样，葛饰北斋似乎自由地采纳线性透视，并在任何他认为合适的时候随意进行调整，从而违反欧洲的透视规则。提到《东海道名所一览》时，安格斯・洛克耶（Angus Lockyer）认为，“葛饰北斋认为无需将自己局限于他能看到的东西，更不必局限于描绘眼前所见。……葛饰北斋毫无拘束地扭曲了这幅地图，以阐明自己的论点。富士山不仅在物质景观中占据空间，而且在精神上提供支柱，日本人的世界由此而构建，观者也可由此找到自己的位置。”[29] 但是，我尤其感兴趣的，是那些在不同表现形式中展现矛盾和对立的例子。以作于 1830—1832 年左右的《富岳三十六景》系列中的《江户日本桥》（图 7）为例，画面上正视图、平行线画法，以及一种类似于线性透视的方式同时可见。斯克里奇对它的解读独具见地：

> 前景中的木桥受到严格的空间限制，并在规模上被锁定在其适当的从属地位，在这里可以看到城市的喧嚣；地位卑微者被横向布置，从而不会受到透视处理的影响。木桥的那一边，城市商业精英们浮华气派的货栈带着其各自的识别标记向远处延伸；这些都在透视规则中展开——这并非异常，因为过度虚饰是商人的本性……不过，在画面的后部，可以清楚辨认出的是大和之域的两个伟大的象征——幕府城堡的角楼和富士山的顶峰。这两个要素不是傲慢或虚荣之物，它们是日本国家和祖先文化崇高的中心枢纽；而正是这些要素依然处于透视体系之外：作为整个宏大政治体系的产物，城堡和山峰被描绘得遥不可及——栖身其下的世俗由一系列平行线联结在一起，但要接近城堡和山峰，任何这些平行线都无济于事。民众被压在下面，民族国家及其宏大的建筑散布在上面；西方透视是支配着中间阶层的机制。[30]

斯克里奇的论述的确站得住脚，而我只想就葛饰北斋对透视本身的运用强调另一点。那些“浮华气派的货栈”，由于既定的观看方向，必须描绘成纵深走向，我们是否可以不假思索地把这种情况看作一个接受透视规则的问题？深度可以通过其他方法表现出来，并非只依靠线性透视。在这个例子中，运河两侧布满临水货栈，沿着运河方向的深度形成了严格意义上的“深远的透视景象”（vista），这些仓库的处理暗示了线性透视；但如图 8 所示，它们远不是“在透视规则中展开”。从这个意义上讲，中间阶层没有真正受西方透视支配，而是在利用西方透视——一种可以开发但又来自异域的观看和理解的方式，从而扮演了与这种特殊方式对话交流的舞台。

还有一批浮世绘版画通常还带有浮绘（*uki-e*）这个词，其真正的舞台描绘，与其说为这种感知方面的对话交流服务，不如说其自身成为一个操纵透视的画面中心；这些作品描绘歌舞伎剧场的娱乐场景，对我们现在进行的讨论有帮助。小林頼子和望月みや引述了奥村政信版画中的一个例子，即《芝居狂言舞台颜见せ大浮绘》，其中天花板的正交线大致汇聚在单一灭点上，而其他正交线在退远时却没有；因此则存在一个所谓的“透视线聚合区域”，以便使舞台相对来说显得更大。小林頼子和望月みや推断，这将吸引潜在的消费者——当他们看着这些版画时，会觉得好像他们就置身在剧院里仔细欣赏着他们最喜欢的演员。[31] 小林頼子和望月みや的注意力继而转向歌川丰春（约 1735—1814 年）创作的一幅作品，认为其透视线聚合区域由于不同灭点的明确分组而变得更加整洁。[32] 确实，我查阅过的大多数歌舞伎剧场版画都表现出一种共同的模式：右侧的正交线大致汇聚在中心稍微偏右的一点上，而左侧的正交线则汇聚在中心稍微偏左的一点上（图 9），这是一种类似于我们称之为水平的“鱼骨”或“透视线聚合轴”的透视画法。

水平方向透视线聚合轴透视，称为洼み绘（*kubomi-e*），也用于室外环境的图像中——建筑物和其他构筑物提供了主要的透视线。小林頼子和望月みや从葛饰北斋的《忠臣藏》中一个场景引用了一个很好的例子。[33] 另一个例子则在此如图 10 所示。葛饰北斋甚至将这种表现方法编绘成一个画面三分原理——他在《北斋漫画》中用两幅插图来说明他所说的“构图三分法”（图 11）。[34] 理论上的程式化并不一定意味着在应用中会严格遵循这些规则。无论如何，即使程式化的透视线聚合轴的透视画法也仍然保持着其撕开所谓统一空间的能力，那么该空间便不再是几何构想和理解中的空间。

* * *

所有这些都发生在江户时代（1615—1868 年），当时日本在很大程度上将自己封闭起来，不受外国的影响；从十七世纪中叶开始，只允许中国人和荷兰人在港口城市长崎与其进行贸易。这意味着，大部分传入日本的西方绘画技术都“经由中国人的视觉翻译而渗入。”[35] 年希尧（1671—1738 年）的《视学》于 1729 年首次出版，1735 年再版。其对日本艺术家真正的影响仍然存在争议，但可以肯定的是，在采用西方绘画传统手法方面，中国在一定时期内一直处于领先。大家公认做到这一点的中国艺术家之一是徐扬，他是苏州人，活跃于十八世纪中叶。1751 年乾隆皇帝第一次南巡时恰好经过苏州，徐扬向乾隆皇帝呈上了他的画作，此后被任命为宫廷画院一等画师。

徐扬特别擅长以挂轴和手卷两种方式描绘广阔的城市场景。让我们首先来看一个挂轴画作的例子——绘于 1767 年的《京师生春诗意图》。正如钟妙昏（Anita Chung）相当谨慎地指出的那样，此画“乍看之下，在安排画面空间上似乎已采用线性透视”，但事实并非如此——宫殿的每个区域及其前面的城市区域的每个部分，都“通过平行、汇聚或发散的线条来单独组织。”[36] 不过这也只是徐扬处理图像手法的一个方面，因为画作中还有许多其他值得注意的“扭曲变形”，其中大多数显然是有意而为。例如，中轴线上的建筑物以及太庙的建筑物都被扩大了，这种方法与人物画中的方法非常一致——社会等级最高的中心人物，与同一幅画中的其他人物相比，其绘制比例通常不同。与正阳门到天安门的距离相比，暗示出的天安门和午门之间的距离增大了，这也许不但为了更加充分展示天安门和午门，而且也可尽现这两座显要门楼前各自相当大的区域。不仅如此，午门到太和门之间的距离愈发扩大，而太和门到太和殿之间距离的夸张尤为显著，以使紫禁城中最重要的庭院格外醒目。徐扬进一步使该庭院微微仰起以达到强化的效果，从而导致其正交线在透视上“偏离”紫禁城其他部分的正交线。所有这些“偏离”和“扭曲变形”，再加上其他没有列举出的细节，使人无法将该画作解读为一个统一且几何上一致的空间。

在徐扬的手卷画作中，这种“偏离”和“扭曲变形”要明显得多——手卷这种体式实际上决定了，不可能在整个画作中营造出统一且几何上一致的空间。[37] 然而情况远不仅于此。以 1759 年创作的《盛世滋生图》为例，在其中去考虑任何灭点都毫无意义。平行线画法通常用于呈现整个画作中建筑物和构筑物，建筑物正面的方向根据其位置或画作布局的需要在偏左和偏右之间交替。如图 12 所示，如遇深度很大的图像——尤其是以锐角向画面内延伸的街道图像——平行线的使用仍然基本上起主导作用，沿街建筑形成一排且向远处延伸，建筑物近大远小，与近大远小的人物形象相称，而这主要是通过依次缩小每座单体建筑来实现的。实际上，在十二世纪初张择端的《清明上河图》（图 13）中就已经可以看到这种手法。其中建筑物由近及远，其尺度的渐小是通过使用平行线和建筑高度递减而实现的。《盛世滋生图》中唯一的例外是那些单体自身连续的建筑物，例如城墙上的城垛、粮仓的外墙、或长廊，其图像使用了汇聚线。在这两个例子中，如果我们认为延伸到远方的特定街道场景的尺度渐小暗示了一个特定的“透视线聚合区域”，那么在手卷任何一个片段——1 到 1.2 米的观看长度——中就会有两个或多个不同的“透视线聚合区域”并存，从而带来严重的空间感知问题。正是由于这种空间问题，郎世宁（1688—1766 年）不得不在其 1728 年创作的《百骏图》中委曲求全地运用带有三个灭点的透视系统，而这一特殊方法，在马可・穆斯洛（Marco Musillo）看来，其实源自“早期现代欧洲的舞台布景传统”（图 14）。[38] 但是对徐扬来说，我们可以推测这根本不是问题。

十年后，徐扬似乎更系统地将平行线与建筑高度从近到远递减结合的这种方法应用到其1770年完成的《乾隆南巡图》中。何慕文（Maxwell K. Hearn）通过与王翚的《康熙南巡图》（作画时间推定为1698年）的比较，正确地指出此手卷中“西方线性透视的改进用法”，不过他同时坚持认为“徐扬遵循西方空间现实主义的惯例”，并“致力于维持一个一致的视点”，我不能肯定这是否完全令人信服，除非何慕文所言“空间现实主义”和“视点”只是大致形象的说法。[39] 我们可以用第六卷《大运河至苏州》中描绘阊门大街的那个片段（图15）来说明我的论点。此处显示的街道是从画面的左下部沿对角线纵深延伸到右侧中部，其尺度沿此方向明显渐小，就好像街道两侧的屋脊、屋檐、露台和屋基构成了一组想象中的连续直线，而这些连续的直线最终似乎可以汇聚。但如图16所示，这些看似连续的直线实际上是由一系列短小并彼此分离的断线形成的，每条断线所描绘的都是单个建筑物的屋脊、屋檐、露台或屋基。表示每座单体建筑的直线基本上彼此平行，只有一、两条断线颇为异常，似乎暗示汇聚。徐扬显然是在利用——甚或从视觉上创造——街道两旁单体建筑之间的分隔，并重新调整每条线的位置或方向，以利于实现尺度渐小的效果，现代人眼中的线条汇聚之感也正因此而生。所以，徐扬在这幅画中并没有真正使用线性透视，也没有设定真正的观看之“点”来巡察他所描绘的城市场景的“空间”。徐扬当然利用了欧洲的再现手法，但我的论点是，他可能对“偏离”这些手法没有任何顾虑，恰恰是因为他像奥村政信和葛饰北斋一样，在心理上或理智上都不受缚于那种空间统一且由几何法则贯穿始终的世界观。

* * *

现代学术研究不断过分关注中国山水画中的线性透视、其准确性、或者缺乏线性透视的问题，其原因恰恰在于把空间当作理所当然的观念。正是基于这条思路，我认为上文已概述的在欧洲和中国绘画之间建立起来的诠释性对比，不仅是一种过于简单的构想，而且是至少有潜在误导性的构想。即使是“投影”（projection）一词及其所表达的概念，诸如“平行投影”（parallel projection）和“轴测投影”（axonometric projection）这样的短语，在诠释现代到来之前的中国山水画时也变得不甚可靠。李约瑟认为中国人“已经意识到了投影问题，即如何在平面上表现三维空间。”[40] 但是，投影不应该等同于三维空间的二维再现，而李约瑟也没有为在中国语境下使用“投影”概念提供明确的限定条件。[41] 土圭的阴影会投射在地面上，而竹子的阴影会投射在园墙上，但是对这些现象的认识和使用，与我们目前在画法几何的理论和技术基础上所理解的“投影”的应用，在概念上是不同的。我们对中国画的研究和欣赏确实已经植根于与其传统不同的另一个参照系中。

确实，古汉语中没有一个表示“空间”（space）的词。[42] 现代汉语中与其对应的词——“空间”——是二十世纪初从现代日语中借来的（或被重新采用的）。由中国表意文字组成的日语词汇“空间”（*kūkan*）和“时间”（*jikan*）创造于明治时期（1868—1912 年），以对应从西方引进的新的概念，并且如道信中贤所说，在日本“直到出现空间（*kūkan*）一词时，空间的意识或概念才在语言上得以规范化。”[43] “空”在古汉语（以及后来的日语）中通常是指我们现在理解的空无一物或空洞，而“间”的繁体字是由“门”和“月”组成的表意文字（閒），许慎（约 30—约 124 年）“定义”它为裂隙或空隙，而徐锴（920—974 年）用比喻的方式评论道：“夫门夜闭，闭而见月光，是有间隙也。”[44] 这个表意字还有另一个不同但又相关的意思：“未占用”，而衍生为“稍稍休闲 / 空闲”。[45] 因此，“间”基本上是指事物或情况的“中间”。

现代的空间概念不仅涵盖我们日常生活的世界，而且包括“宇宙”（universe, cosmos），与后者接近的现代之前的中国概念用“天地”和“宇宙”表达。“天地”在翻译上似乎没有“宇宙”更具争议。《庄子·庚桑楚》一节中有两个句子，用“宇”和“宙”解释“道”。华兹生（Burton

Watson）是这样翻译的：

> It has reality yet there is no place where it resides—this refers to the dimension of space. It has duration but no beginning or end—this refers to the dimension of time.[46]
> （译文：有实在而无其处所——这指的是空间的维度。有持续而无始终——这指的是时间的维度。）
> （原文：有實而無乎處者，宇也；有長而無本剽者，宙也。）

我不得不怀疑，在“宇”和“宙”上分别强加“空间的维度”和“时间的维度”这两个现代概念或许既帮助我们理解原文，也导致我们误解原文。但是问题仍然在于，如果不是“空间”和“时间”的“维度”，那应该是什么的“维度”，对此我承认我目前还没有答案。《淮南子·齐俗训》中有两个与《庄子》之句不同而具共鸣的句子，安德鲁·迈耶（Andrew Meyer）的翻译如下：

> From furthest antiquity to the present days is called “extension-in-time”;
> The four directions [plus] up and down are called “extension-in-space.”[47]
> （译文：从最久远的古时到现今，这叫做“时间中的延伸”；四方［加上］上与下，这叫做“空间中的延伸”。）
> （原文：往古來今謂之宙，四方上下謂之宇。）

尽管此译文试图帮助现代的人们理解近两千年前撰写的文本，但仍给我带来类似的不安。与中国文学传统中大多数概念的情况一样，在这些引用的语境中这两个表意文字具有很强烈的比喻特点，远离了任何抽象定义的意味。在《淮南子·览冥训》中，同是这两个字也曾组合起来使用，这一次是指其原始含义——建筑的屋檐和屋脊。[48] 不幸的是，在比喻性陈述上强加现代人熟悉的抽象概念，有扭曲原作者对世界的构想和观察的趋势。我的观点是，无论从经验的维度还是宇宙的维度来讲，现代之前的中国文本传统中都没有我们现在所理解的“空间”，而我们在这一传统中所面对的是一方没有“空间”的天地，而山水与园林就处在这方天地之中。

从宇宙这一最大规模转到我们日常生活的范围，我们会更清楚地意识到这个概念已经深深植根于现代人的头脑中。翻译中习惯或先验地使用“空间”的情况很多，在此从一些备受尊敬的学者的著述中选一些例子进行简要说明就足够了。例如，宣立敦（Richard E. Strassberg）在翻译欧阳修（1007—1072 年）的《醉翁亭记》（1046 年）时，为其中著名的一句精心选词：“But wine is not uppermost in the Old Drunkard’s mind. What he cares about is to be *amid* mountains and streams”（原文：醉翁之意不在酒 在乎山水之間也）。宣立敦对于文化差异的确非常敏感，但在翻译白居易（772—846 年）的《三游洞序》（819 年）时，他却把“兩崖相廞間”翻译为“in the *space* between the two cliffs”。除了“相廞”二字在译文中漏译之外，尽管“space”（空间）一词或许不太必要，但它的出现并没有带来太大问题，因为其意思可以是“宽度”、“间隔”或“距离”。但是，当这个词出现在柳宗元（773—819 年）《钴鉧潭西小丘记》（809 年）中的一段时，情况就完全不同了：

> I lay down using a mat as a pillow. Clear and cool forms sought out my eyes, the gurgling sound of water sought out my ears, *the expansive space* sought out my spirit, and the capacious quietude sought out my mind.[49]
> （斜体字译文：广袤的空间）
> （原文：枕蓆而臥 則清泠之狀與目謀 瀯瀯之聲與耳謀 悠然而虛者與神謀 淵然而靜者與心謀。）

在现代感知模式中，那“广袤的空间”与人的精神进行交流，这当然说得通，但是柳宗元会感觉与其精神交流的真是这个空间吗？

在中国艺术史上，“空间”的使用非常频繁。例如高居翰（James Cahill）将郑思肖（1241—1318年）印在《墨兰图卷》（1306年）上的闲章中的词句翻译如下：

> If you ask me, you won't get it; if you don't ask, I may give it to you. In the *great space and expanse* before my eyes, there's only the pure breeze of the present and the past.
> （译文：如果你向我提出请求，你不会得到它；如果你不提出请求，我或许会把它给你。我眼前那广大的空间和辽阔的区域中，只有现下和往昔的清风。）
> （原文：求則不得 不求或與 老眼空闊 清風萬古。）

正如高居翰强调的，这是“任意翻译的”。的确，这“任意性”尤其反映在高居翰对原文前两句的诠释和理解：这两句似乎揭示了郑思肖对于生活和自然法则的理解，而非指“你”、“我”之间进行交流的情况；其“任意性”也反映在高居翰漏译郑思肖的“老眼”，而“老眼”对于突显最后四个字的力度至关重要。不过这里“space”一词的使用显然既颇欠准确，又具有误导性，而从高居翰通篇明了顺畅的讨论来判断，任意翻译的可能性极小。[50]

这个问题不论在任意翻译的文字还是在严格翻译的文字中同样存在。这里仅举一个例子，即以不同方式和理解诠释郭若虛（十一世纪下半叶）的《图画见闻志》（约1080年）中的名句“深遠透空 一去百斜。”[51] 亚历山大·科本·索伯（Alexander Coburn Soper）的翻译版本是：

> Perspective distances will penetrate the space, with a hundred [lines] converging on a single point.[52]
> （译文：透视距离将穿透空间，一百[条线]汇聚在一个单点上。）

线性透视被当作解读十一世纪中国学者言语的指导原则，这种解读所暗示的是文艺复兴后的欧洲绘画传统普遍适用于所有文化。卜寿珊（Susan Bush）和时学颜（Hsio-yen Shih）的翻译看起来是在疏离线性透视：

> Deep distances penetrate into space and a hundred diagonals recede from a single point.[53]
> （译文：深远的距离穿透空间，而一百条斜线从一单点退远。）

这里出现的是一个似是而非的情况：卜寿珊和时学颜的翻译制造出内在矛盾，恰恰是因为移除了任何对线性透视的明显引征，但“空间”和“单点”仍然以一种唤起线性透视的方式被保留下来，从而导致后半句现在变得毫无意义。

罗伯特·J·前田（Robert J. Maeda）周密地将此句与其前面的句子相结合，并使他的翻译没有任何线性透视的意味：

> ... and brushstrokes of even strength should deeply penetrate space, receding in a hundred diagonal lines.[54]
> （译文：……强度均匀的笔触应深深穿透空间，以一百条斜线退远。）

前田对这句话的解释的确很有见地。不过无论是否暗示着线性透视，在上面引用的例子中，“空”字都被转换为“space”（空间）。但是我们知道，“空”并不意味着“空间”，而且我怀疑这些杰出的学者也了解这一点，至少当这一特定的汉字成为他们研究的中心议题时应该如此。正是基于这种思路，我认为卡斯韦尔的翻译明智地摆脱了透视或空间的负担：

> Deep distance penetrates the *voids* and together yields many slanting or dispersed [views].[55]
> （译文：深远的距离会穿透虚空，共同产生许多倾斜或分散的[可观察到的景象]。）

翻译中类似的趋势也展现在园林方面的学术著作中。邓肯·M·坎贝尔是这样翻译李斗（1749—1817 年）的《扬州画舫录》中一段文字的："Touring this *space* one feels oneself to be like an ant crawling through the twisting eye of a pearl, ..."[56]。（译文：游历这个空间，人会觉得自己就像一只蚂蚁在珍珠中曲折的孔眼中爬行。原文：遊其間者如蟻穿九曲珠。）坎贝尔素以其令人羡慕的流畅翻译技能著称，但如前面已经简短阐述，"间"字在现代之前的汉语文献中既没有、也绝不暗示"空间"的意思。最后再举一个例子，即李惠仪（Wai-Yee Li）翻译的祁彪佳（1602—1645 年）的《寓山注》，其中《宛转环》同样包含这样的句子："Here one starts walking and in no time reaches the summit of the Distant Lodge (Yuange 远阁): this is Old Man Gourd's (Hugong 壶公) magical art of contracting *space*."（译文：从这里，人们开始步行，须臾而至远阁的顶层：这是壶公缩小空间之魔术。原文：兹纔一舉步 趾已及遠閣之巔 是壺公之縮地也。）[57] 这里，"地"应该被更恰当地解读为"地方"或"区域"，将它翻译成"space"在某种程度上不仅曲解了原意，而且也使其丧失了可能的社会和体验方面的言外之意。

这些选自受人尊敬的学者著作的例子，当然无意损害其令人钦佩的成就。我本人最初受过建筑学方面的训练，一直坚定地行走在现代空间概念的轨道上，而且我以前的著述中也充斥着"空间"一词；但是使用我自己的例子远没有引用这些学者的例子更有说服力。因此，我在此仅说明"空间"

概念在我们对中国绘画、园林和山水的感知和思考中有多么根深蒂固、多么普遍。甚至连郭熙（约 1020—约 1090 年）在其《林泉高致》中构想的著名的“三远”图式，有时也被认为是空间感知的方法。但是，事实是“三远”只明确涉及山脉。正是出于这个原因，甚至当这一基本模式被著名的、具有极高准确性和敏感性的艺术史学家方闻（Wen Fong）稍加扩展，成为“感知景观的方式”，我们或会为此警觉。[58]

人们可能仍然会从另一个角度为此争辩，就像米特罗维奇（Mitrović）在方法论上警告我们时所强调的，不应“将分析中使用的概念与这些概念应该分析的假设相混淆”，并且还坚持认为“以理论概念解释某种现象一般滞后于这个现象的出现，不能将理论概念的一时缺乏当作是人们对此现象毫无意识。”[59] 但是，“空间”的概念并不仅仅是在分析中使用的；它经常构架、塑造我们的感知。另一方面，如果李成和其十世纪同时代画家已经意识到“空间”现象，但需要经过约一千年绘画传统的持续发展、直到大约在二十世纪初用以解释这一现象的概念才突然出现；这一概念的特定术语又必须从现代日语借用；而该术语仅在十九世纪后期才在现代日语中创造出来——这样的情况几乎不可能存在。

* * *

与我们现代空间概念有关的古汉语中的字词，除了上面已经提到的那些之外，还有很多。其中有些可以在英语中找到或远或近的对应，而另一些则找不到。这些术语大多数表示或暗含位置、方向、情况和相互关系。其中大部分都反映出两极概念（向 / 背、高 / 下、远 / 近、狭 / 阔、虚 / 实，等等），无论一对字词中的另一个在特定语境下是否出现。即使是“广袤”这个词，经常被翻译成一个笼统的英语单词或短语，例如“expanse”或“great expanse”，实际上也是两极化的，因为“广”是指东西方向和距离，而“袤”则指南北方向和距离。中国文明是地球上现存历时最久的文明，在历史上一直充满生机，而非停滞不前，做到这一点的前提是历史的延续与变化，而这种延续和变化结合的特点，借用李克曼（Pierre Ryckmans）用典于帕门尼德斯（Parmenides）和赫拉克利特（Heraclitus）时的妙语：是一个“持久并不否定变化，而是塑造变化”的过程。[60] 这种连续性也体现在自汉至清末发展而来的对世界的感知方式。

要证明文化中某个现象的普遍缺乏，收集再多的证据也无济于事。因此举出“什么不是”的例子仅仅是为了说明一个观点。例如，《周礼·考工记》中关于典范城市营建形态的著名段落是：

> 匠人營國 方九里 旁三門 國中九經九緯
> 經涂九軌 左祖右社 前朝後市 [61]

这面这段文字给我们留下深刻印象的是格局、数量、方向和位置的相互关系，而没有任何空间的暗示。

这种强调方向和位置相互关系的地图描述方式看来一直持续发展，在北宋（960—1127 年）之后的游记和园记中都继续使用，尽管一方面其表现形式更为丰富详尽，另一方面则结合了新的描述方式，即游历方式。[62] 在这方面，朱长文（1039—1098 年）的《乐圃记》是园记中一个颇为贴切的例子。[63] 该园林的布局以地图和游历两种描述方式的巧妙组合而呈现，将读者的注意力从一个区域转移到另一个区域：首先简要提到以堂三楹统领的居住区域；由此进入到居住区域以南的一个独特区域，其特色是三开间的遂经堂；然后，从遂经堂的西北角把读者引入见山冈和其下的池塘；该小山之北又为西圃；最后在西圃的西南出现了另一座自然小丘，名为西丘。因此，整座园林的呈现是一个五个区域以方向和位置上的相互关系为线索相继展开的过程，而除最后一个区域外，每个区域依次以某重要的人工结构或突出的自然地貌为中心，这个中心则成为描述该区域内各部分之间相互的方向和位置关系的参照点，偶尔出现的两个区域之间的相互参照也只是为了表达某种思想，而非空间定义。毫不奇怪，这些方向和位置的相互关系都没有被量化。

在建筑形象与平面结合的地图中，方向和位置相互关系在中国感知模式中的首要地位尤其明显。以 1229 年的《平江图》为例，它与现代到来之前的中国绝大多数城市地图一样，采用的再现手法似乎将每个建筑群和其他由社会或文化界定的区域“分解”开来。但这与我们所了解的“立体分解图”根本不同。“立体分解图”的物体呈现在遵循几何规则、有固定视点、并在统一的空间中合乎逻辑地分开各个组成部分。对比之下，《平江图》则呈现出众多系列的展开区域。如图中的子城（图 17）所示，我们看到每个区域中的建筑正面图像仰面平铺在游访者的前方、左侧和右侧，暗示想象中的游访者从一个区域移动到另一个区域；大型门道（如子城南大门和西侧门）的双扇门保持半开状态，而中轴线上连接小堂、宅堂以及后罩房的廊屋屋面则看起来好像从上俯视而得。图 18 中的提刑司和周围街道的形象更加说明问题：提刑司建筑群的入口在东端，第一个院落的北、南、西三面每栋建筑的正面图像仰卧，恰似来访者在庭院中所见；但来到第二个庭院时，南侧建筑亦为仰卧，但相对于来访者的位置而言却上下反向，这就意味着该建筑就使用而言应该不是提刑司建筑群的一部分，而其正面和入口应该朝南，面向建筑群外面的街道。《平江图》中的另一个例子来自两个小型院落建筑群，即居养院和威果四十一营，位于南城墙脚下的南星桥西面（图 19）。两个院落的入口都在北面，彼此内部建筑的朝向却有所不同：居养院的布局与《平江图》中大多数院落一样，简单明了，但是在营房的图像中，院落内部的建筑都为南向，这很容易理解——这些建筑是士兵起居之所，不论出于生理还是文化原因，它们应该朝南，而居养院中的建物并非日常的人类居所，因此不需要这种常规的朝向。

图中假想的游访者预期会遵循适当的或一般的路线，其视线和视野基本保持在形象与平面结合的地图中，建筑的方向和位置关系则根据游访者沿着路线而移动的位置来呈现。屋面的情况是个例外，其图像所暗示的观者从地图之外的位置远距离观看屋面；换句话说，观者在这类地图中看到的情况很大程度上是地图内部假想的游访者的视线和视野。然而，在园林绘画中，可见的游访者成为景观的一部分；观者在绘画中看到的情况不同于图像中游访者的视线和视野。钱贡（活跃期 1579—1616 年）所绘的版刻手卷《环翠堂园景图》为我们提供了一个很好的例子。图 20 包含的几个片段描绘了四个相邻区域，即“环翠堂”的前院、“兰亭遗胜”、“冲天泉”和带有“百鹤楼” 的“达生台”。在这里，即使是未经训练的眼睛也应注意所有这四个区域之间“空间”上既不一致也不谐调。这些区域彼此间的一致性仅限于其相对位置和由墙壁、台阶、入口建立的相互联系，以及建筑和人物的比例、花草树木的位置。[64] 这些区域展现在我们眼前，但我们无法确定任何一个特定视点；也就是说，视点和线性透视的概念，以及平行“投影”概念，都与这种感知模式完全无关。简而言之，无限的、各向同性和各处同质的空间概念不但在其中无处容身，而且与之无益。

我不否认现代空间概念作为工具在分析中国园林和山水方面的价值，事实证明现代空间概念对于我们思考现代建筑和园林设计尤其有用。我甚至并不是在提议讨论现代之前的中国艺术、建筑、园林等时避免使用“空间”一词。我只是建议，如果我们试图了解现代之前的中国园林和山水画是如何被其创作者构想出来并被其观赏者所感知的，那么重要的是要记住，这种现代概念在那些创造者和观赏者的头脑中根本不存在。既然人的感知与人的思维基本相符，那么正如马西莫・斯科拉里（Massimo Scolari）在讨论耶稣教士试图将透视引入中国的尝试中所强调的那样，“改变观看的方式，亦即改变再现的方式，意味着改变思维模式”。[65] 将现代概念强加在现代之前的中国观念和感知上，我们实际上也就更改了他们的思维方式，曲解则因此而来。约翰・海伊（John Hay）以中国人在图像上或文字上是否“提及空间这个话题”为开篇，对“中国绘画中的中国空间”进行了精彩的讨论，但我认为这里其实谈不上是否“提及”此话题。相反，它是关于概念的存在本身和思维方式的特殊性的问题。话说回来，我非常重视海伊缜密地对其讨论对象所提出的限制条件：即为了讨论方便起见，他强调“空间只是延伸，是任何类型的事物组成所必然暗示的。换句话说，那是存在赖以发生的条件。”[66] 在现代到来之前，中国事物组成的类型肯定不同于欧洲事物组成的类型。在中国语境中，这种“延伸”没有必要同性同质、无穷无尽地延续，而“存在赖以发生的条件”也并不一定要以具体尺寸和统摄划一为前提。那些书画家和造园者胸中之丘壑构成的是独特的、不受几何空间限制的一方自成体系的天地。

1

Figure 1 : Okumura Masanobu, *Enjoying the Cool of the Evening at Ryôgoku Bridge, an Original Perspective Print*, ca. 1745, woodblock print, 34.3 × 47 cm. Museum of Fine Arts Boston. Public-domain image

图 1：奥村政信，《两国桥夕凉见 浮绘根元》，约 1745 年，木版画，34.3 厘米 × 47 厘米。波士顿美术馆。公共领域图片

2

Figure 2 : Okumura Masanobu, *Elegant Female Version of Shutendoji, an Original Perspective Print*, ca. 1745, woodblock print, 29.9 × 43.1 cm. Museum of Fine Arts Boston. Public-domain image

图 2：奥村政信，《风雅女酒吞童子 浮绘根元》，约 1745 年，木版画，29.9 厘米 × 43.1 厘米。波士顿美术馆。公共领域图片

3

Figure 3 : Okumura Masanobu, *Kabuki Theatre District in Sakai-chô and Fukiya-chô, a Large Perspective Picture*ca. 1745, woodblock print, 43.8 × 64.5 cm. Museum of Fine Arts Boston. Public-domain image
图 3：奥村政信，《境町葺屋町芝居町 大浮绘》，约 1745 年，木版画，43.8 厘米 × 64.5 厘米。波士顿美术馆。公共领域图片

4

Figure 4 : Okumura Masanobu, *Elegant Shakuhachi Version of Ushiwakamaru Serenading Jôruri-hime, an Original Perspective Picture*, 1740s, woodblock print, 32.1 × 45.7 cm. Museum of Fine Arts Boston. Public-domain image

图 4 : 奥村政信，《风雅尺八十二段 浮绘根元》，十八世纪四十年代，木版画，32.1 厘米 × 45.7 厘米。波士顿美术馆。公共领域图片

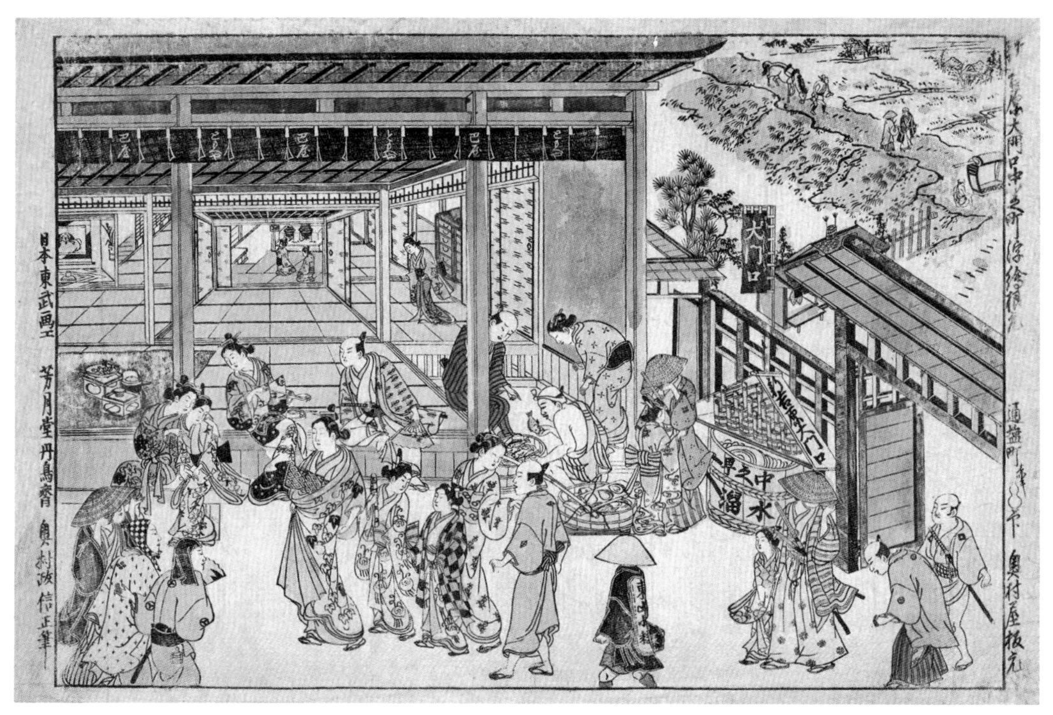

5

Figure 5 : Okumura Masanobu, *Great Gate and Naka-no-chô in the Shin Yoshiwara, an Original Perspective Picture*, 1740s, woodblock print, 32.6 × 46.3 cm. Museum of Fine Arts Boston. Public-domain image
图 5 : 奥村政信，《新吉原大门口中之町 浮绘根元》，十八世纪四十年代，木版画，32.6 厘米 × 46.3 厘米。波士顿美术馆。公共领域图片

Figure 6 : Okumura Masanobu, *Interior of Echigo-ya in Suruga-chô, a Large Perspective Picture*, ca. 1745, woodblock print, 43.7 × 62.2 cm. Museum of Fine Arts Boston. Public-domain image

图 6：奥村政信，《骏河町越后屋吴服店 大浮绘》，约 1745 年，木版画，43.7 厘米 × 62.2 厘米。波士顿美术馆。公共领域图片

7

Figure 7 : Katsushika Hokusai, *Nihon-bashi in Edo* in the series *Thirty-Six Views of Mount. Fuji*, ca. 1830–32, woodblock print, 25.4 × 36.8 cm. The Metropolitan Museum of Art. Public-domain image
图 7 : 葛饰北斋，《富岳三十六景》系列中的《江户日本桥》，约 1830—1832 年，木版画，25.4 厘米 × 36.8 厘米。大都会艺术博物馆。公共领域图片

8

Figure 8 : Katsushika Hokusai, *Nihon-bashi in Edo,* with guiding lines in red added to show accurately the utilisation of linear perspective
图 8 : 葛饰北斋，《江户日本桥》，加上红色透视线以说明线性透视在此图像中的运用方式

9

Figure 9 : Utagawa Toyoharu, *Perspective Picture: Depiction of a Kabuki Theatre*, ca. 1776, woodblock print, 26 × 37.7 cm. Art Institute Chicago. Public-domain image
图9：歌川丰春，《浮绘歌舞伎芝居之图》，约1776年，木版画，26厘米 × 37.7厘米。芝加哥艺术馆。公共领域图片

10

Figure 10 : Katsushika Hokusai, Act IX of *The Newly Published Perspective Picture of the Loyal Retainers*, 1801–1804, woodblock print. The British Museum. Public-domain image
图 10：葛饰北斋，《新版浮绘忠臣藏》第九景，1801—04 年，木版画。大英博物馆。公共领域图片

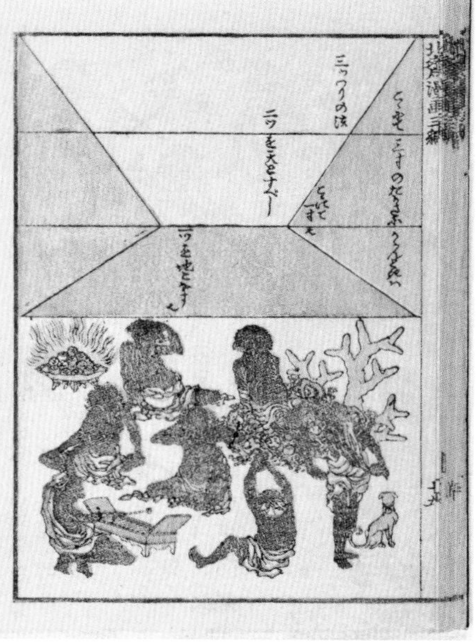

11

Figure 11 : Katsushika Hokusai, *Hokusai Manga*, 3.15b–16a, 1815, woodblock print. The British Museum. Public-domain image
图 11：葛饰北斋，《北斋漫画》，3.15b—16a，1815 年，木版画。大英博物馆。公共领域图片

12

Figure 12 : Xu Yang, *Burgeoning Life in the Prosperous Age,* 1759, handscroll, ink and colour on silk, 35.8 × 1225 cm. The section depicts the Ten-Thousand-Year Bridge and the Provincial Surveillance Offices where the provincial examination is in progress. Liaoning Provincial Museum. Public-domain image
图 12：徐扬，《盛世滋生图》，1759 年，手卷，绢本水墨设色，35.8 厘米 × 1225 厘米。此片段描绘正在举行省府考试的万年桥和臬台衙门。辽宁省博物馆。公共领域图片

13

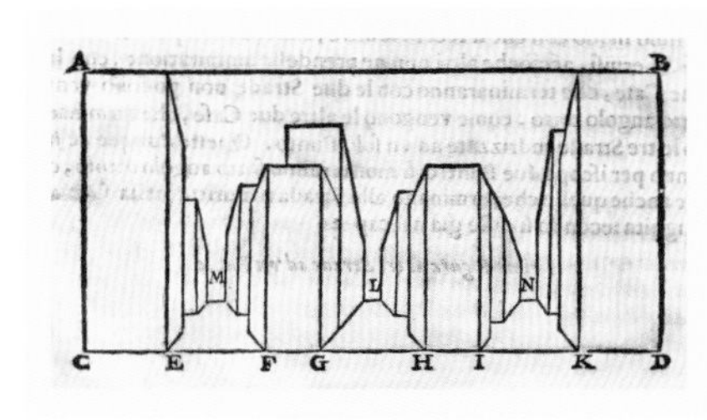

14

Figure 13 : Zhang Zeduan, *Along the River at the Qingming Festival* (detail), early 12th century, handscroll, ink and colour on silk, 25.5 × 525 cm. Palace Museum, Beijing. Public-domain image
图 13 : 张择端，《清明上河图》（细部），十二世纪早期，手卷，绢本水墨设色，25.5 厘米 × 525 厘米。北京故宫博物院。公共领域图片

Figure 14 : Nicola Sabbattini (1574–1654), *Practica di fabricar scene, e machine ne' teatri*, 1638, page 48
图 14 : 尼古拉・萨巴提尼（Nicola Sabbattini，1574—1654 年）《舞台布景与机械结构手册》，1638 年，第 48 页

15

16

Figure 15 : Xu Yang, *The Qianlong Emperor's Southern Inspection Tour*, Scroll Six: Entering Suzhou along the Grand Canal (detail), 1770, handscroll, ink and colour on silk, 68.8 × 1994 cm. The section depicts the Changmen Street. The Metropolitan Museum of Art. Public-domain image
图 15：徐扬，《乾隆南巡图》，第六卷《大运河至苏州》（细部），1770 年，手卷，绢本水墨设色，68.8 厘米 × 1994 厘米。此片段描绘阊门大街。大都会艺术博物馆。公共领域图片

Figure 16 : Xu Yang, *The Qianlong Emperor's Southern Inspection Tour,* Scroll Six: Entering Suzhou along the Grand Canal (detail), with guiding lines in red added to show accurately the utilisation or non-utilisation of linear perspective
图 16：徐扬，《乾隆南巡图》，第六卷《大运河至苏州》（细部），添加了红色透视线以准确地说明线性透视在此图像中的运用或非运用方式

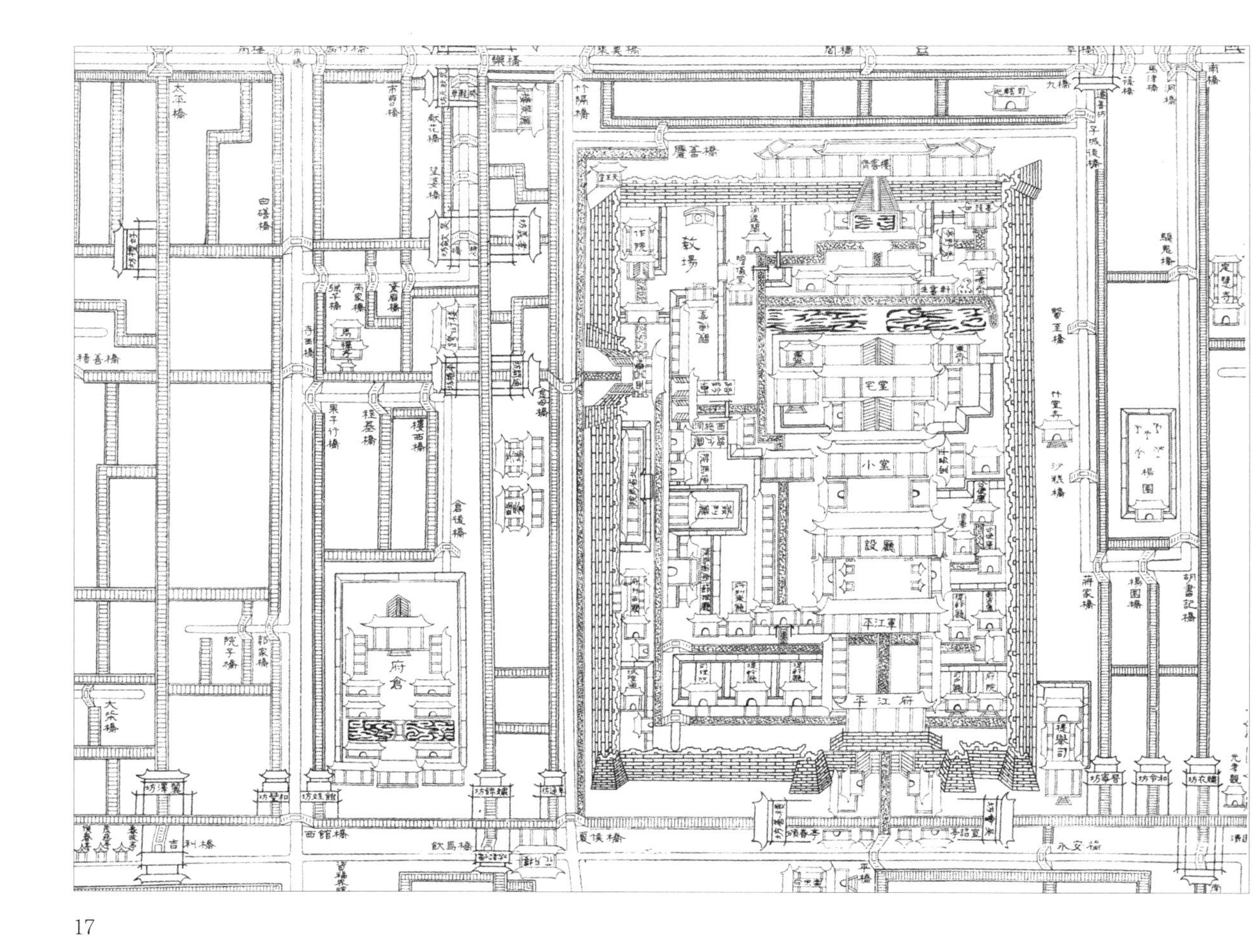

17

Figure 17 : Line drawing of detail of the *Picture-map of Pingjiang*, carved on stone in 1229. The section depicts the inner walled enceinte of the city. Courtesy of the Suzhou Committee of Urban and Rural Construction, for this image and those in Figures 18–19
图 17：据 1229 年《平江图》碑所制的线描图（细部）。此片段描绘平江城子城。感谢苏州城乡建设委员会提供此图以及图 18–图 19 中的图片

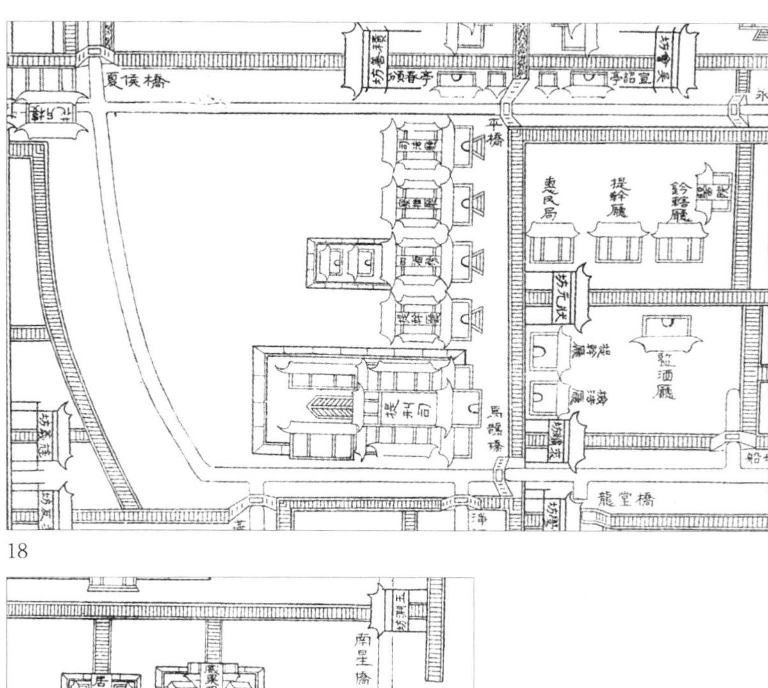

18

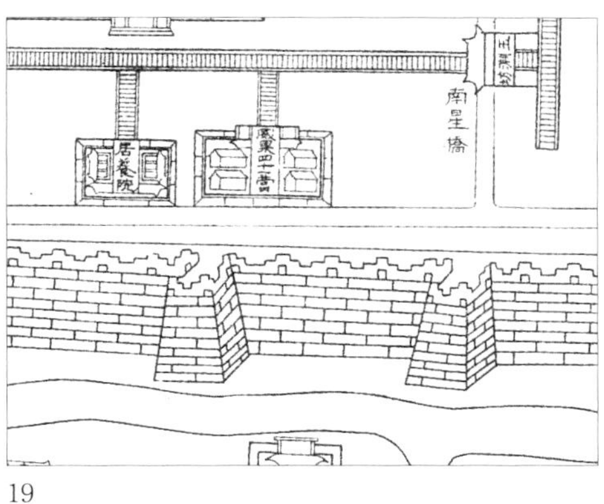

19

Figure 18 : Line drawing of detail of the *Picture-map of Pingjiang,* depicting the Justice Office and surrounding streets
图 18 :《平江图》碑线描图（细部），描绘提刑司和其周围街道

Figure 19 : Line drawing of detail of the *Picture-map of Pingjiang*, depicting the Juyang Almshouse and the 41st Weiguo Barracks, to the west of the Southern Star Bridge at the foot of the south wall of the city
图 19 :《平江图》碑线描图（细部），描绘南城墙脚下南星桥西面的居养院和威果四十一营

20

Figure 20 : Qiangong, *Illustrations of the Hall Surrounded by Jade* (detail), handscroll, 24 × 1470 cm, facsimile copy (Beijing: Renmin meishu chubanshe, 1981), plates 27–29
图 20：钱贡，《环翠堂园景图》（细部），手卷，24 厘米 × 1470 厘米，影印本（北京：人民美术出版社，1981 年），图版 27–29

园林六则提要

葛明

1

“园林六则”来自我对园林的认识，是从中发展起来的设计方法之一。对于园林研究而言，一些当代中国建筑师更属意于山水画传统中的“业余画家”——宋代苏东坡那一类历史文人的方法，希望“取其大意”。其中王澍、董豫赣、童明老师等和我一样都属于对此心向往之的同类研究者，都愿意把研究的方向之一集中为“园林的方法”。

我们所说的“园林的方法”不止是一种在当代设计园林的方法，更多地指向如何参与到当代建筑学之中，是试图以园林作为媒介、工具、能动机制的方法。需要说明的是，本文语境内的园林大部分是指中国园林。我们希望“园林的方法”能够加深对建筑学的理解，同时对专业的园林研究者们也有所裨益。

表 1：设计方法研究分类

空间的方法 / Method of Space
- 体积法 / Raumplan
- 结构法 / Structure-ing
- 不定形法 / De-formal Planning

类型学 / Typology

概念建筑 / Conceptual Architecture

园林的方法 / Method from Garden

“园林的方法”是当代研究设计方法的分支之一。表1是我对所研究相关设计方法的大致分类，包括空间的方法、类型学、概念建筑的方法等等，其中之一即“园林的方法”。它位置特殊，一方面需要尽可能地与其他方法不同或拉开距离，另一方面又需要和其他方法进行对应才易于理解和发展。阿尔多·罗西表述过类似的意思，画画、舞蹈等都是从模仿开始的，而建筑是从不模仿开始的，因此而成为建筑学，它体现理性。这一表述的特殊意义在于为建筑学的学科性做了铺垫。然而园林需要从模仿自然开始，它本身体现了人们对待模仿的一种态度，但这一模仿与画画、舞蹈又有所不同，它需要有某种特殊的综合的想象力，这就是“园林的方法”和通常的罗西式的建筑学所不同的地方——它从模仿开始，而模仿的核心是对历史的想象和对自然的想象的结合。

关于历史想象与自然想象的结合，可以以元代著名书画家赵孟頫的《鹊华秋色图》（图1）为代表。在赵孟頫的时代，他希望摆脱南宋的绘画风格，在绘画中带有古意或者是韵味。这一名作特意舍弃了南宋传统，所有画面里的物体都变得简拙起来。这是一种特意追求的拙，在拙的物体之间形成了特殊的空，它们又一起形成了整体的拙，让我们对自然和对历史交错在一起的情形产生了新的认识。这一似乎难以琢磨的方法，充分体现了园林方法的一种理想——通过对历史的想象和自然想象并置、交叠甚至特意制造矛盾从而产生意义。我所研究的园林方法的大意就深受这一幅画的启发，这也是园林与画结合思考的起点之一。

此外，我模仿中国的一些传统，试图为园林的方法通过画而设立三个梯阶，这三个梯阶如同学习书法和绘画所设的梯阶一样。这些梯阶表明作为“园林的方法”，它试图寻求可以循序渐进的道路，这也是作为方法来理解的园林的特殊性，与许多研究者所认为的园林研究不应有如此强烈的分阶而不同。所设的三阶分别以龚贤、王蒙、郭熙的画为范本，尤其后者可以用鬼神莫测来形容其状态。

园林方法三阶的第一阶以龚贤的作品为例（图2），画中体现了两个特征——物体的突出；物体与空的并重。第二阶以元代王蒙的《青卞隐居图》（图3）为例，画中所有的物体和空已全部交错在一起，难以区隔，甚至连空都已经被排挤出去，只余物体。第三阶以郭熙的《早春图》（图4）为例，画中充满了对物体的奇特描述，以及对物体之间空的奇特描述，它们互相错杂、交叠在一起，又共同形成了对氛围的奇特描述。以上分别体现了对园林方法三阶各自状态的想象。

园林方法三阶有着共同的特征：特殊状态的“远”、特殊的物体与空、特殊的交错，所有这些构成了鲜明的特征，与通常所认为的园林往往以消除物体而提供场地感、提供氛围的方法不同。以上说明，我所在意的“园林的方法”并不在意完全交融在一起的状态，它寻求物体和空之间是否能达成一种莫名状态的连接。在这种莫名的状态中，对山水画中的“远”就有可能获得新的认识。在此，“远”并非目的，“远”试图同时使物体和空呈现。

2

在此基础上，我试图简要提出园林方法的一种——园林六则。六则的前三则分别是现-生活模式之变、型、万物。其中，万物包含山石第一、中介物第二、房屋花鸟池鱼再次之、林木无定所。万物的核心之一是要有生气的万物，万物的另外一个核心是要让物体感突出，从所有事物中凸显出来，才能被经营。第四则是坡法，可理解为倾斜的要素的方法，中间包含了结构、材料等等。第五则是起势，它与名词不同。最后一则是真假，其词性为形容词，提供了美学上的重要判断。

以下是对园林六则的部分解释。例如第二则是型，它既是类型，又是形式。型最重要的是如何使自然与人工的世界互成。以计成的《园冶》为例，《园冶》中第一篇就是相地——

如何寻找地、观察地。在“相地篇”中一共有六种地，其一为山林地，其二为城市地，其三为村庄地，其四为郊野地，其五为傍宅地，其六是江湖地。在我看来每一种地就是一种园林的类型，每一种地都有其处理边界的方式，每一种地都预示着在地上来建一个园林，既要融合于地，但是要区别于这种地，既在画面之中又在画面中显现出来，如此才是所谓的型。而各类园林之间难的是在一个自然的地里经营园林，因此，山林地园林或许是最复杂的园林。

为描述型，试以型中的“水土之法”为例。在龚贤的一幅画（图5）中，水面、水边的棚屋、堤岸、水面、植物堤岸依次排开，似乎表现了一种平远，但是平远又不断地被打断，出现了另外无法用高远或者深远来描述的空间，这就是上文所提及的物体和物体之间的空，它们以一种神奇的方式作用在一起，互相作为背景，互相显现，互相渗透，显示出一种强烈的透明性。这一特征若从画面来看就是水土之法中的水土相隔，利用水土相隔制造了特殊的远，制造了特殊的势，从而使在这种远、这种势之上放置任何一处房子的时候，都可以产生一种既属于此又不属于此的状态，这就是运用了水土之法这一型所产生的结果。

同理，在龚贤的另一幅画（图6）中可以看到在堤岸上的房子，好像既属于堤岸，又深陷于水中，但是似乎又在堤岸河水中间限定出了新的地方，同样，这一房子促成了水土相隔。因此，水土相隔既是水和土的直接相隔，也是利用其他物体来进行相隔的一种方法。

水土相隔在现实中的令人称颂的例子有无锡的寄畅园，还有京都的西芳寺，即苔寺。它有个迷一样的平面——号称心形的平面（图7），水和土互相相隔，其平面的状态与通常所说的一池三岛不同，岛、水的分量巧妙均衡，形成了迷离的效果。我们无可置疑地能感受得到这似乎是一片自然地，但同时，专业人员也不难发现，这不是一片普通的自然地，水土相隔经过了非常仔细的均衡，最终呈现了一个野中有文的园林，或者说是文野相间的园林，类似做法在传说中的宋代、明代的一些园林中间不乏先例。

为进一步解释型，我还试图引入一个特殊的术语“中介物”来展开讨论，仍以龚贤的画（图 8）加以说明。画中，在水平展开的堤岸和垂直耸起的高峰之间有一房子，房子的屋顶之下由一片空构成，这片空好像有人曾待过又离开了，而这就是另一种隔，是中介物的一种，它使得水平和垂直用一种空交接起来，使画面产生了戏剧性的变化。这处变化和远山以及近处的房子交融在一起，呈现出一个看上去非常自然的场景，而靠近眼前的又是另一番表演。它们交叠在一起，而所有这些都依赖于这一小小的房子中间这一空白。所以，用最小的空白使得自然和不自然的东西接在一起，这就是中介物所能起到的作用。而设计一个园林，就似乎总是在设计或者在寻找这类中介物，用最小的力气达成“野”和“文”之间的联系。

3

如何发展园林的方法？我一直尝试以系列设计作为研究，从中不断体会到园林和画这一议题对于发展方法的潜能。

首先是由一组厂房改造而成的微园（图 9-图 15），我尝试运用园林六则中的型和万物这两则。设计希望使原有厂房变成充满空的一个地方，所用的办法是在水平上形成很多间隔。白色的坑、黑色的地板、白色的梁架、绿地、黑色墙面，层层地展开，不断地切分，产生了不断的分隔，形成了特殊的远。建设时期的微园西厅照片清晰地表达了这种隔——寻找特殊中介物来处理内和外，以及或内或外的状态。

其次是太湖边上的四个小园子设计（图 16），我试图表达出山水、园林、画、物体、空之间复杂的往来关系。其呈现的方式首先是关于龚贤画的类比图（图 17），接着是模型（图 18），表达了四组非常特殊的物体。我关注它们如何既融于环境，又能凸显出来，并以这种方式构筑园的认识。在此模仿龚贤画图的方式和模型的方式是同样重要的。

画、物体互相交叠，这一思维方式——在画和物体之间故意模糊的思维方式，都体现了对待园林和画的关系的一种认识。

其中已建成的春园（图 19-图 27）用地窄而长，用作游客中心。我试图采用“半园半房”这一“型”——以房中有园、园中含房的方式回应场地，实现舒展的同时又能获得曲折尽致。在方法上则“架构”与“分地”并举。

其三是在徐州正在营建的水徐园局部（图 28-图 32），其模型和图纸再次试图表达隔如何形成实体与空之间来回的交叠。这些场景中有着强烈的物体感，但在中心地带，都显现了隔的作用。

对以上这些对园林方法的研究，做一个简要小结：我先假设在中国的山水画中存在三种状态，分别以龚贤、王蒙和郭熙的画为例，从中阅读出奇特的物体和空交织的方式，这种阅读帮助我产生了对型、对中介物、对水土相隔的认识，从而帮助我确定了园林的方法的一些认识基础。

或许，这是对山水画的一种特意误读，但是中国的园林绝大部分已成为传说，我们能以什么方法来学习园林，并从中提取对当代的意义呢？是否能有效地或者故意地“误读”这些与之相关的画，从而接近传说中的那些园林？在我看来，如何结合自然的想象和历史的想象，如何连接画和园林，从而推动对园林的理解是一种重要的方法。

A Brief Introduction on Six Approaches to Garden

GE Ming

(translation by GE Ming, GU Kai)

1

"Six approaches to garden" is one of the design methodologies I have developed from my understanding of gardens. Some Chinese contemporary architects prefer the methods of "Amateur painter" in traditional landscape painting—like Su Shi in the Song Dynasty and other historical scholars in garden studies who yearned for "understanding the essence". In the same vein, Professor Wang Shu, Dong Yugan, Tong Ming, and others, together with me, are all dedicated to similar research methods, preferring to make the "Method from Garden" one of our study focuses.

The so-called "Method from Garden" is not just a contemporary method of garden design. It is more about how to engage in contemporary architecture using the garden as a medium, a tool, and an agency. Notably, the garden in this text refers mainly to the Chinese garden. We hope that the "Method from Garden" can be conducive to deepening the understanding of architecture, along with being of benefit to professional garden researchers.

The "Method from Garden" is a branch of contemporary design methodology. Table 1 is a general classification of the design methods I have developed, including the method of space, typology, and conceptual architecture. Among the "Method from Garden" occupies a special position. On the one hand, it should be intentionally differentiated and separated from other methods. On the other hand, it must correspond to those methods

Table 1 : Design Methods Classification

空间的方法 / Method of Space
- 体积法 / Raumplan
- 结构法 / Structure-ing
- 不定形法 / De-formal Planning

类型学 / Typology
概念建筑 / Conceptual Architecture
园林的方法 / Method fom Garden

to be well understood and developed. As Aldo Rossi puts it, architecture does not begin with imitation, while painting and dancing do. Hence architecture can become a discipline that is rationally controlled. This statement is significant in that it paves the way to understanding the discipline of architecture. A garden, however, is an expression of people's attitude to nature, and as such, it necessarily begins with an imitation of nature. Yet, this form of imitation is different from those in painting and dancing because the imitation in garden design requires a special and comprehensive imagination. This is how the "Method from Garden" differs from what Rossi regards as architecture. It begins with imitation, but its core is the combination of the imagination of history and nature.

When it comes to the combination of historical and natural imagination, the *Vibrant Autumn Scenes the Ch'iao and Hua Mountains* by Zhao Mengfu (Figure 1), a famous calligrapher and painter in the Yuan Dynasty, can be a representative example. During the era of Zhao Mengfu, his work sought break away from the painting styles of the Southern Song Dynasty and endow his paintings with a kind of ancient style or charm. In this painting, the tradition of the Southern Song Dynasty was deliberately abandoned, with all objects becoming simple and clumsy. The deliberate pursuit of simplicity and clumsiness formed a special relationship between objects and void/emptiness. It created a novel understanding of the interweaving of nature and history. This perceivable yet indescribable method demonstrates the ideal of "Method from Garden," whereby meaning is created through juxtaposition, overlapping, and even deliberately contradicting historical and natural imagination. The essence of "Method from Garden" which I propose is deeply inspired by this painting, and it is also where my thinking of combining garden and painting begins.

In accordance with some Chinese traditions, I would like to set three levels within the "Method from Garden" by paintings. Just like levels of progression in learning calligraphy and painting, these three levels show a gradual progression in the "Method from Garden". This is the particularity of understanding garden through the lens of a method which differs from the general idea that garden research should not have such a significant hierarchy. The three levels here take paintings of Gong Xian, Wang Meng, and Guo Xi as models, the last of which stands for a state, which is the acme of perfection/unpredictability.

The first level takes Gong Xian's painting (Figure2) as an example. This picture manifests two features—the highlighting of objects, and the equal emphasis on objects and void/ empytiness. The second level is exemplified bythe *Dwelling in Qing and Bian Mountains* by Wang Meng in the Yuan Dynasty (Figure 3). All the objects in this painting are completely interlaced with the void/emptiness and are difficult to separate from it. In empty areas the empty space has been squeezed out, leaving the objects only. The third level, demonstrated by the Early Spring by Guo Xi (Figure 4), is filled with peculiar descriptions of objects, as well as unusual descriptions of void/ emptiness among objects—intricately interwoven with each other to create an extraordinary description of ambiance. These examples give evidence of what we mean by the imagination within the three levels of "Method from Garden".

The three levels described above share some distinctive features: a "distance" in a specific state, specific objects and emptiness, as well as specific interlacing. These features depart from the common idea that a garden provides a sense of place and atmosphere by means of clearing away the sense of object. In the "Method from Garden" what matters is not a state of complete blending, but a strange articulation between objects and emptiness. In this elusive state of articulation, we may acquire a new understanding of what is meant by "distance" in landscape paintings. However, this "distance" is not the goal itself, but rather an attempt to present the objects and void/emptiness simultaneously. Otherwise, it is no different from distance in nature, and can never become the intersection of the "distance" in nature and that in the mind.

2

Based on the above, I will briefly propose the six approaches to the "Method from Garden". The first three of the six approaches include "the changing of mode of living", "type-form", and "myriad living things". Among them, "myriad living things" contain mountain rocks to be first; intermediate things to be second; buildings, flowers and birds, ponds and fish to be following; and trees having no fixed place. One of the core aspects of this approach is the vitality ensured for the objects in its application. Another important aspect of this approach is to highlight the sense of objects, for things can only be arranged when they stand out from everything else. The fourth approach the "*po-fa*" can be understood as the method of inclined elements, including structure, material, etc.. The fifth approach is "generating," which, unlike the nouns, indicates the motion. The sixth and final approach is "real/false"—a pair of adjectives significant in aesthetic judgments.

The following paragraphs give further explanation of one of the six approaches. The "type-form" approach refers to both type and form. The most important aspect of this approach is how to make nature and the artificial world complement each other. The first part of the Craft of Gardens is the so-called "Land Seeking"—looking for and observing the land. There are altogether six kinds of land in the chapter titled "Looking for / Observing Site." These are: mountain-site land, urban-site land, village-site land, country-site land, villa-site land, and lake-site land. Each type of land may be interpreted as a type of garden, with its corresponding way of managing the border. Thus, in each type of land lurks a potential garden, blended with the land but distinguished from it—implicated in the picture but standing out from it. This is the so-called "type-form." It is most difficult to create a garden in a natural land, and the mountain-site gardens are probably the most complex among different types of gardens.

To give an example of the "type-form" approach, I would like to describe the "method of water and land." In a painting by Gong Xian (Figure 5), the water surface, waterside huts, embankments, water surface, and plants are arranged in turn, depicting the "level-distance". However, the "level-distance" is interrupted by another kind of space which cannot be described as "high distance" or "deep distance". This is the void/emptiness among objects that I described earlier. They set each other off as background and infiltrate each other, demonstrating strong transparency. The feature here is the *ge* (separation through creating distance/partition) of water and land in the picture. The *ge* of water and land creates the special "Distance" and special "*Shi*"(tendency) that sets the house in an ambiguous state of simultaneous belonging and not belonging. This is the result of the application of the"type-form" of "method of water and land."

Similarly, in another painting by Gong Xian (Figure 6), we can see a house by the bank. Seeming to reside on the bank and sink into the water simultaneously, it defines a new place between the bank and river and contributes to the *ge* of water and land. Therefore, besides their direct separation, the *ge* of water and land could be achieved through the use of other objects.

Some remarkable examples of *ge* of water and land in real gardens are the Jichang Garden in Wuxi and the Saihoji Temple in Kyoto. The latter has an amazing plan—known as the heart-shaped plan (Figure 7), with the *ge* of water and land. The state of the plan is different from what is commonly referred to as one pool and three islands, where the island and water accomplish subtle balance to form a mystical effect. Undoubtedly, we can feel it as a natural land; but at the same time, it is not hard for professionals to see that this is not an ordinary natural land, where the *ge* of water and land has been very carefully balanced, resulting in a wild garden integrated with cultivation, or a garden interweaving wilderness and cultivation. Similar practices also appear in quite a few legendary gardens in the Song and Ming Dynasties.

To further explain the "type-form", I'd like to introduce a special terminology, which "intermediate things". Take Gong Xian's painting for example again (Figure 8). There is a house between the horizontally expanded embankments and the vertically rising peaks. Under the roof of the house is an empty space, where someone may have stayed and left. This is another kind of *ge*, a kind of intermediate things that integrates the horizontal and vertical with empty space, creating a dramatic change in the picture. The change blends with the distant mountains and the nearby houses, creating a seemingly natural scene, which becomes theatrical as it comes close to the eyes. The two scenes overlap, hinging on the empty space in the little house. This is the way in which the intermediate thing acts: as the smallest empty space that brings the natural and unnatural together. The design of a garden seems always to incorporate and seek after such intermediate things to accomplish the link between the "wilderness" and the "cultivation" with the least effort.

3

In developing the "Method from Garden" I always try to research by serial designs, through which I gradually increase my understanding of the potential of the topics of garden and painting to develop the method. The following paragraphs give examples of this research in practice.

The first is Weiyuan Garden, a renovation of factory buildings (Figure 9–15). The design intends to turn two existing buildings into a place full of void/emptiness. The method adopted was to form a lot of *ge* horizontally. White pits, black floors, white beams, green land, and black walls spread out in layers, segmented constantly to create constant *ge*, resulting in a special sense of "Distance." The photo of the west hall in construction in Weiyuan Garden also clearly expresses the *ge*/partition. It seeks special intermediate things to deal with the internal and external, or an ambiguous state between the internal and external.

What follows is the design of four small gardens by the Taihu Lake (Figure 16), in which I attempt to show the complicated interrelation between Shan-Shui, garden, drawing, object, and void/emptiness. It is presented at firstly in the analogy of Gong Xian'painting (Figure 17), then the models (Figure 18), which are actually four groups of highly special objects. My concern is the way they stand out while blending in the environment, which structures the understanding of the garden. Herein, the imitation of Gong Xian's painting and the method of modeling are equally important. The painting and objects overlap each other. This way of thinking intentionally blurs the boundary between painting and objects, expressing a unique understanding of the relationship between garden and painting.

One of the four gardens, Spring Garden, has already been constructed (Figure19–27). In order to respond to the narrow and elongated site, I adopted a "half garden-half-house" type-form in the design, with part of the garden in the house and part of the house in the garden. Thereby, the construction can be both outward and inward. In terms of the methods employed, "structuring" and "separate-joining" were used simultaneously.

The third is the Shui Xu Garden to be constructed in Xuzhou (Figure 28–32). Again, its drawings and model try to demonstrate the overlapping of objects and void/emptiness formed by the *ge*. The scenes have a strong sense of objects, the central places of which all embody the *ge*.

To briefly summarize my research. I assume that there are three states present in Chinese landscape paintings, as shown in the paintings by Gong Xian, Wang Meng, and Guo Xi. And I identified peculiar methods of interweaving objects and void/emptiness in those paintings. This way of reading landscape paintings allows me to gain an understanding of "type-form", "intermediate things", and "the *ge* of water and land", and thus helps me identify some cognitive basis for the "Method from Garden".

This may be a deliberate misinterpretation of landscape painting. However, when most Chinese gardens have become legends, how could we study gardens and extract their contemporary significance? Is it possible for us to approach those legendary gardens by effectively or intentionally misinterpreting the related paintings? In my opinion, it is an important method to combine natural and historical imaginations and to articulate paintings and gardens, which promotes the understanding of gardens.

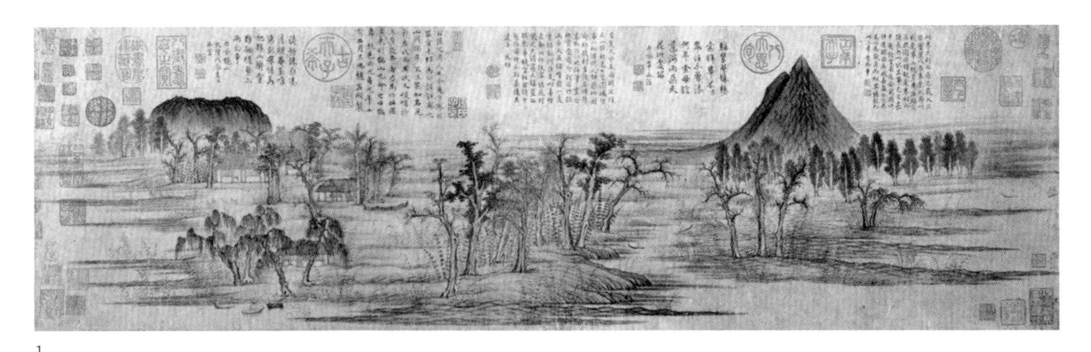

1

2

图 1：历史想象与自然想象 我的园起 / （元）赵孟頫《鹊华秋色图》
Figure 1 : Zhao Mengfu (1254–1322, Yuan Dynasty), *Vibrant Autumn Scenes the Ch' iao and Hua Mountains* / the imagination of history and nature

图 2：（明清）龚贤《八景山水卷》
Figure 2 : Gong Xian (1618–1689, late Ming and early Qing Dynasties), *Eight Paintings of Landscapes*

图 3：（元）王蒙《青卞隐居图》
Figure 3 : Wang Meng (1308–1385, Yuan Dynasty) *Dwelling in the Qing and Bian Mountains*

3

4

图 4：（北宋）郭熙《早春图》
Figure 4 : Guo Xi (1000–1087, Northern Song Dynasty), *Early Spring*

5

6

图 5：（明清）龚贤《山水册》
Figure 5 : Gong Xian (1618–1689, late Ming and early Qing Dynasties), *Album of Landscapes*

图 6：（明清）龚贤《山水册》
Figure 6 : Gong Xian (1618–1689, late Ming and early Qing Dynasties), *Album of Landscapes*

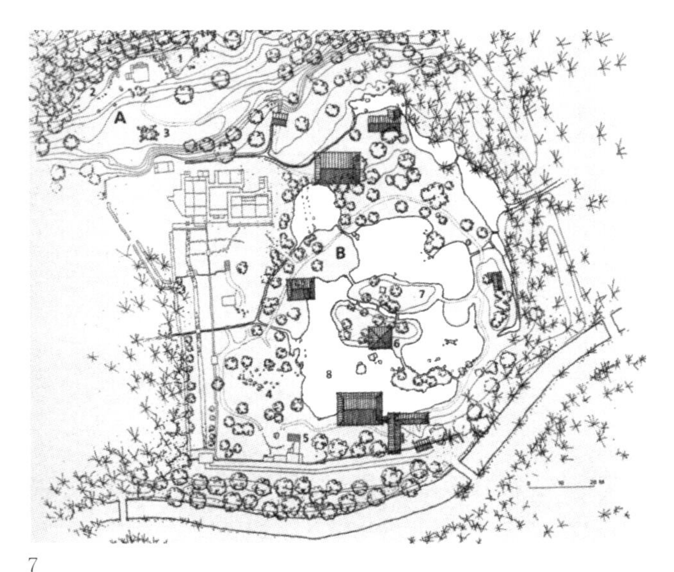

7

8

图 7 : 京都《西芳寺》
Figure 7 : Saihoji Temple, Kyoto

图 8 : （明清）龚贤《山水册》
Figure 8 : Gong Xian (1618–1689, late Ming and early Qing Dynasties, *Album of Landscapes*

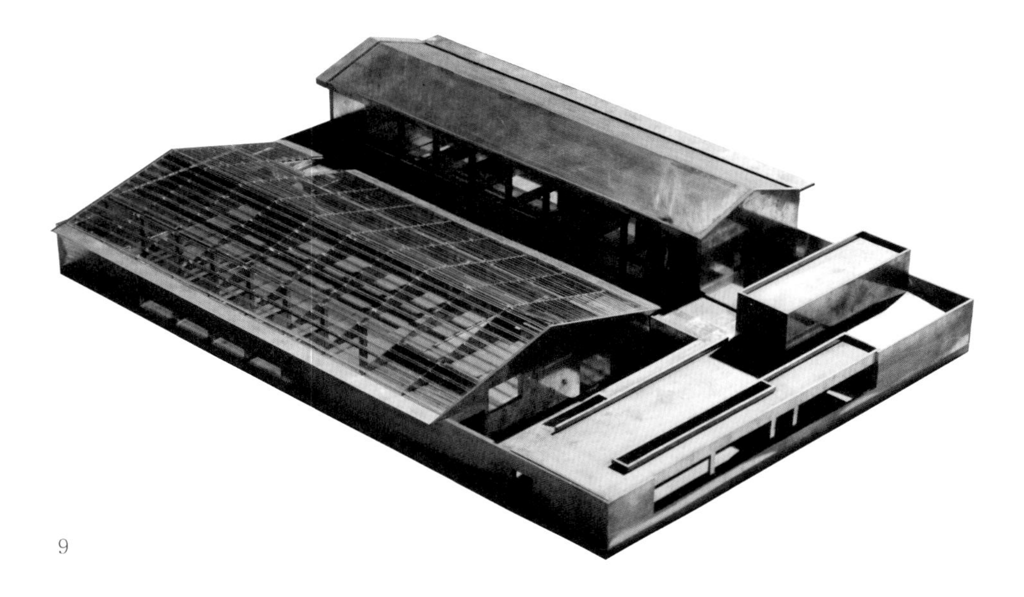

9

图 9 : 微园模型
Figure 9 : Model of Weiyuan Garden

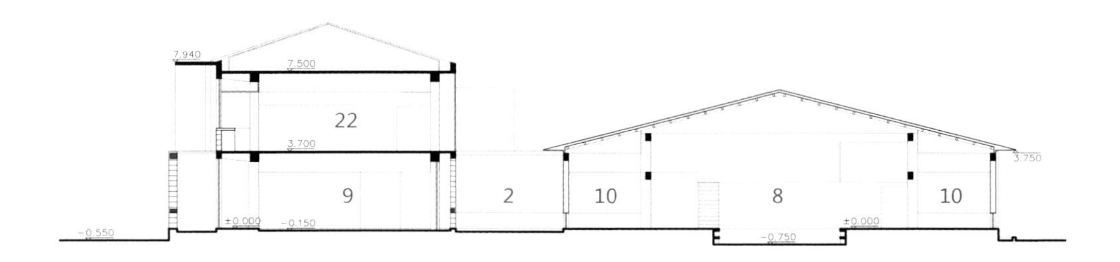

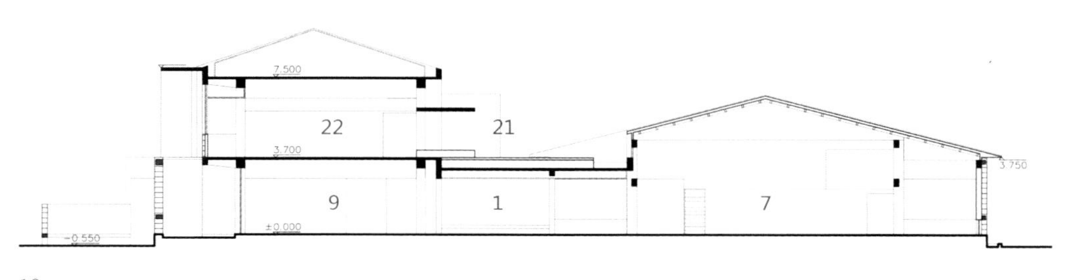

10

1 门厅 Entrance Hall
2 中院 Centre Yard
3 南院 South Yard
4 北院 North Yard
5 墨池 Ink Pond
6 四面厅 Full View Pavilion
7 白厅 White Hall
8 书池 Book Pond
9 黑厅 Black Hall
10 书法展厅 Exhibition Room
11 存储 Storage
12 复厅 Repeated Hall
13 读书处 Reading Room
14 竹院 Bamboo Yard
15 厨房 Kitchen
16 厂区配电间 Existing Distribution Room
17 配电间 Distribution Room
18 书法表演 Calligraphy Studio
19 看山台 Mountain Terrace
20 观水台 Water Terrace
21 听风台 Wind Terrace
22 办公室 Office Room

图 10 : 微园剖面图
Figure 10 : Sections of Weiyuan Garden

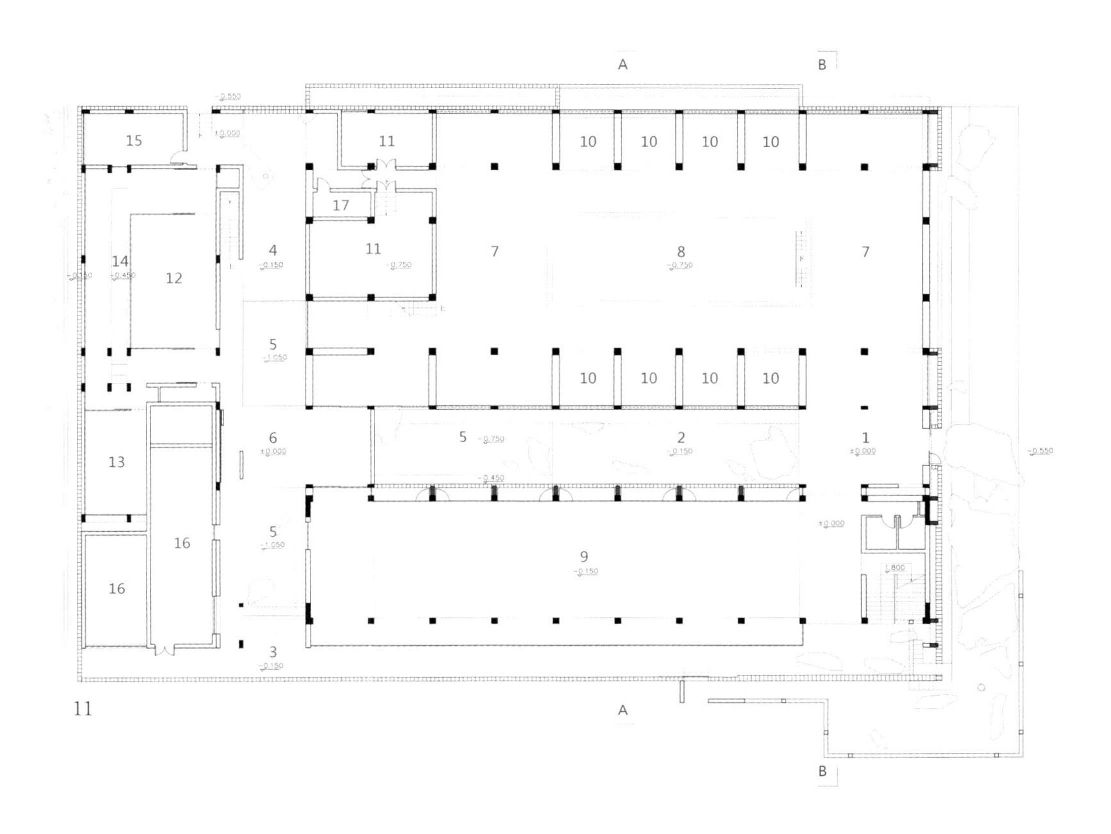

图 11 : 微园平面图
Figure 11 : Plan of Weiyuan Garden

12

13

14

图 12 : 白厅。赖自力拍摄
Figure 12 : White Hall. Photo by Lai Zili

图 13 : 复厅。孔德钟拍摄
Figure 13 : Double Hall. Photo by Kong Dezhong

图 14 : 施工中的复厅。孔德钟拍摄
Figure 14 : Double Hall under construction. Photo by Kong Dezhong

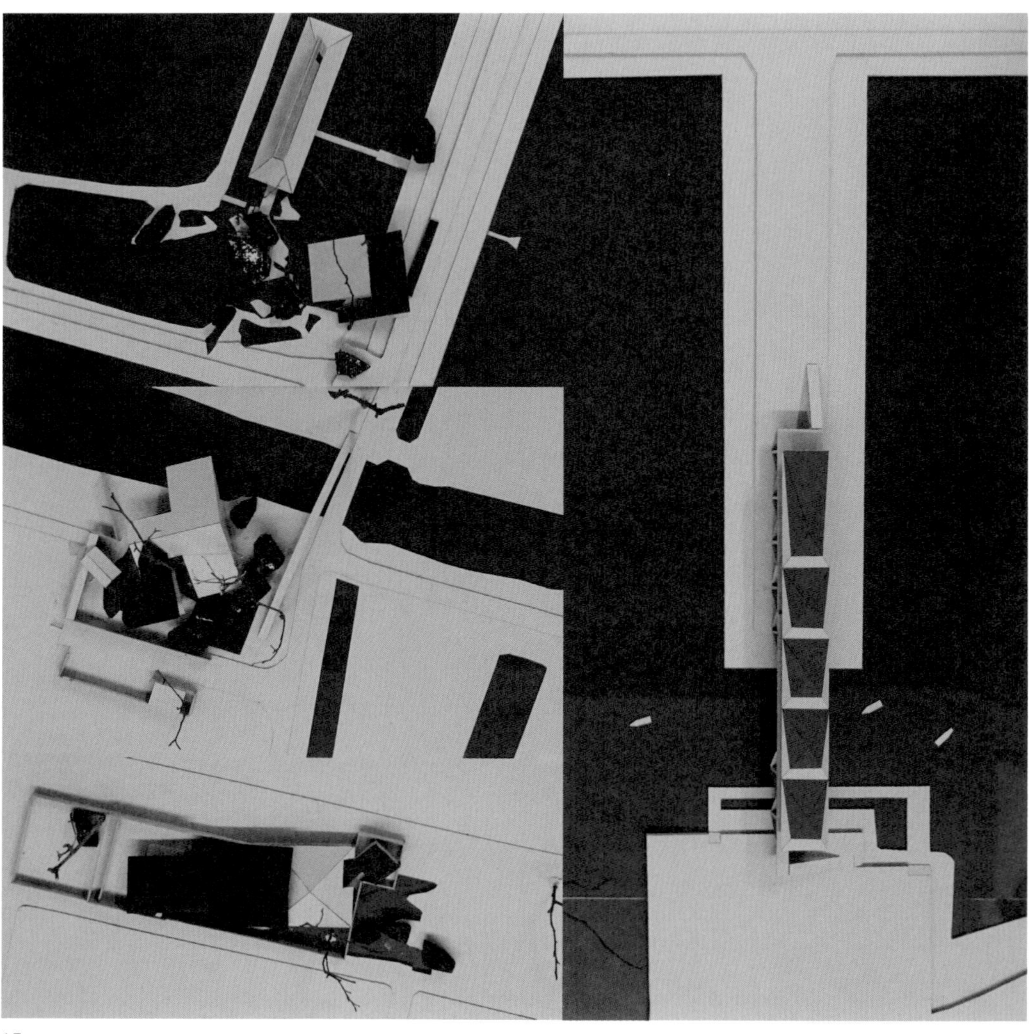

15

图 15：三园一市模型
Figure 15 : Model of four Gardens by the Taihu Lake

16

图 16 : 龚贤画类比图
Figure 16 : Analogy painting of Gong Xian

17

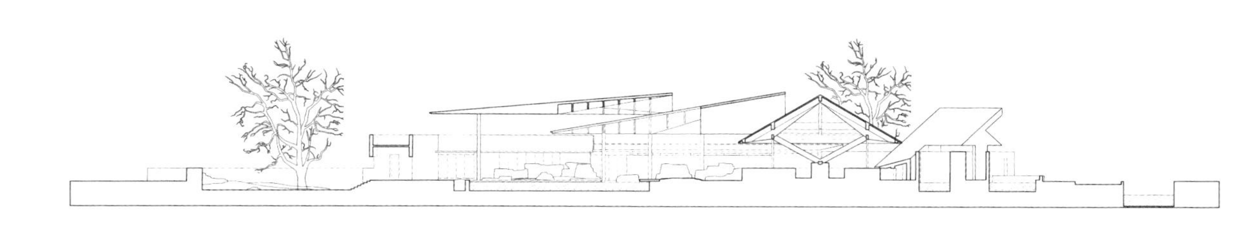

18

图 17 : 春园模型
Figure 17 : Model of Spring Garden

图 18 : 春园剖面图
Figure 18 : Sections of Spring Garden

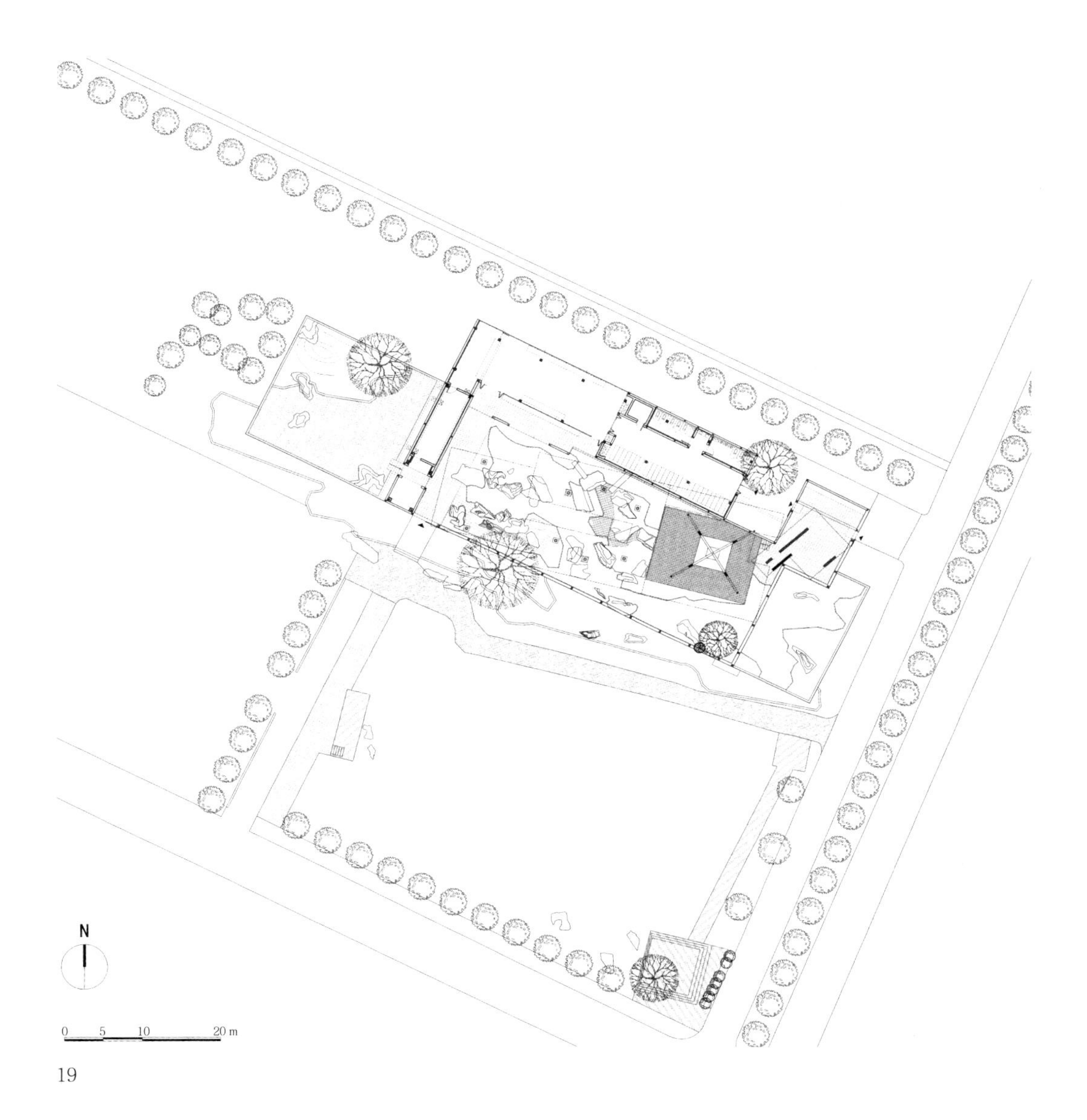

19

图 19 : 春园平面图
Figure 19 : Plan of Spring Garden

20

21

图 20 : 南望春园。陈灏拍摄
Figure 20 : South facade. Photo by Chen Hao

图 21 : 西望春园。陈灏拍摄
Figure 21 : East facade. Photo by Chen Hao

22

23

图 22 : 沙园。陈灏拍摄
Figure 22 : Sand Garden. Photo by Chen Hao

图 23 : 苔园 1。陈灏拍摄
Figure 23 : Moss Garden 1. Photo by Chen Hao

24

图 24 : 苔园 2。陈灏拍摄
Figure 24 : Moss Garden 2. Photo by Chen Hao

25

图 25 : 方台。蒋梦麟拍摄
Figure 25 : Platform. Photo by Jiang Menglin

26

图 26 : 徐州水徐园局部模型之一
Figure 26 : One part of model of Shui Xu Garden in Xuzhou 1

27

图 27 : 徐州水徐园局部模型之二
Figure 27 : One part of model of Shui Xu Garden in Xuzhou 2

28

图 28 : 徐州水徐园场景之一
Figure 28 : One part of model of Shui Xu Garden in Xuzhou 1

29

图 29 : 徐州水徐园场景之二
Figure 29 : One part of model of Shui Xu Garden in Xuzhou 2

NOTES

Introduction

1 A useful anthology of Euro-American texts on Chinese gardens is *Ideas of Chinese Gardens: Western Accounts, 1300–1860*, ed. Bianca Maria Rinaldi (Philadelphia, PA: University of Pennsylvania Press, 2016).

2 For a lucid account of these changes during the period of disunity, see Donald Holzman, *Landscape Appreciation in Ancient and Early Medieval China: The Birth of Landscape Poetry*, contained in idem, *Chinese Literature in Transition from Antiquity to the Middle Ages* (Aldershot, England: Ashgate Publishing Co., 1998).

3 See, for instance, Zhou Weiquan, "Wei Jin Nanbeichao yuanlin gaishu," in *Jianzhushi lunwenji di liu ji*, ed. Tsinghua daxue jianzhuxi (Beijing: Tsinghua University Press, 1984), 80–95; idem, *Zhongguo gudian yuanlin shi*, 2nd ed. (Beijing: Tsinghua daxue chubanshe, 1999), 81–120; Wang Yi, *Yuanlin yu Zhongguo wenhua* (Shanghai: Shanghai renmin chubanshe, 1990), 76–108.

4 Also translated as "dreaming journey" by some other scholars.

5 Shen Yue (441–513), *Song Shu* (Beijing: Zhonghua shuju, 1974), 93.2279; Zhang Yanyuan, *Lidai minghua ji*, ann. Yu Jianhua (Shanghai: Shanghai renmin meishu chubanshe, 1964), 129.

6 Guo Xi (ca. 1000–ca. 1087) and Guo Si (*jinshi* 1082), *Linquan gaozhi*, in *Meishu congshu*, ed. Huang Binhong and Deng Shi (Hangzhou: Zhejiang renmin meishu chubanshe, 2013), 2.7.6. My translation of the passage follows Wen Fong, "Monumental Landscape Painting," in *Possessing the Past: Treasures from the National Palace Museum, Taipei*, ed. Wen C. Fong and James C.Y. Watt (New York: The Metropolitan Museum of Art, 1996), 131.

7 Osvald Sirén, *Gardens of China* (New York: Ronald Press Company, 1949), 4. Sirén then immediately qualifies this suggestion:"One must not, of course, press the comparison between the landscape paintings and the garden compositions too hard, for it is a matter of two different art forms in which the respective mediums of expression differ widely, but the comparison is nonetheless valuable, and it serves to illustrate the attitude of the Chinese toward problems of composition."

8 I here borrow the beautiful wording from F.W. Mote,"A Millennium of Chinese Urban History: Form, Time, and Space Concepts in Soochow," *Rice University Studies* 59, no. 4 (Fall 1973): 39, where he states: "we are concerned with a vital China, not a moribund one."

9 Craig Clunas, *Fruitful Sites: Garden Culture in Ming Dynasty China* (London: Reaktion Books, 1996).

10 The term *huayi* is usually translated as "ideas in painting" or "picture-idea," but I am inclined to follow the argument made by John Hay, "Boundaries and Surfaces of Self and Desire in Yuan Painting," in *Boundaries in China*, ed. idem. (London: Reaktion Books, 1994), 162: the character *yi* "is frequently more an issue of 'intent,' involving a directionality of feeling and, quite clearly, the vectors of desire." See also idem, "Some Questions Concerning Classicism in Relation to Chinese Art," *Art Journal* 47, issue 1 (1988): 32.

11 Kristina Kleutghen, "From Science to Art: The Evolution of Linear Perspective in Eighteenth-Century Chinese Art,"in *Qing Encounters: Artistic Exchanges between China and the West*, ed. Petra Ten-Doesschat Chu and Ning Ding (Los Angeles, CA: The Getty Research Institute, 2015), 173.

Living amid the Mountains and Waters of the Six Dynasties

1 *Jinling tujing*, in Yue Shi (930–1007), *Taiping huanyu ji, juan* 90, in *Riben Gongneiting Shulingbu (Kunai-chō Shoryō-bu) cang Song Yuan ban Hanji xuankan*, ed., An Pingqiu et al. (Shanghai: Shanghai guji chubanshe, 2013), 58.375.

2 Chen Liang (1143–1194);"Wushen zai shang Xiaozong Huangdi shu,"1188, contained in Chen Yi (1469–1538), *Jinling gujin tukao* (Nanjing: Nanjing chubanshe, 2006), 74.

3 Fan Ye (398–445), *Hou Han shu* (Beijing: Zhonghua shuju, 1965), 83.2756.

4 Chen Shou (233–297), *Sanguo zhi* (Beijing: Zhonghua shuju, 1959), 48.1158.

5 Liu Yiqing (403–444), *Shishuo xinyu* (Shanghai: Shanghai guji chubanshe, 1982), 2.31.22.

6 Ibid., 2.73.32.

7 Ibid., Chapter 4: "On Literary Learning" (*wenxue*) contains more than sixty entries on idle talk, on the basis of which I made the summary here.

8 Ibid., 4.55.19.

9 Ibid., 4.61.29:"Jian Wen entered the Hualin Garden, looking around, and said:"Distance matters nothing if you have it in your heart. He saw lush woods and enjoyed his leisured time, feeling relative to birds, animals and fish."

10 Xu Song (fl. mid-8th century), *Jiankang shilu*, written in 756 (Beijing: Zhonghua shuju, 1986), 12.444.

11 He Shangzhi (382–460), "Hualin Qingshudian fu," contained in *Yiwen leiju*, ed. Ouyang Xun (557–641) (Shanghai: Shanghai guji chubanshe, 1965), 62.1125.

12 Wang Rong (467–493), "Sanyu sanri qushuishi xu," contained in *Yiwen leiju*, 4.72.

13 Fang Xuanling (579–648), et al., *Jin shu* (Beijing: Zhonghua shuju, 1974), 79.2075–76.

14 Li Yanshou (fl. the middle 7th century), *Nanshi* (Beijing: Zhonghua shuju, 1975), *juan* 75 and 76 list 31 hermits, and *juan* 71, 19 Confucian scholars.

15 Fang Xuanling et al., *Jinshu*, 94.2440.

16 Xiao Zixian (487–537), *Nan Qi shu* (Beijing: Zhonghua shuju, 1972), 22.414; Yao Silian (557–637), Chen shu (Beijing: Zhonghua shuju, 1972), 25.321; Shen Yue (441–513), *Song shu* (Beijing: Zhonghua shuju, 1974), 67.1754.

17 Zuo Si (ca. 250–305), "Zhaoyin," contained in *Zhaoming wenxuan*, ed. Xiao Tong (501–531) et al. (Zhengzhou: Zhongzhou guji chubanshe, 1990), 22.297.

18 Shen Yue, *Song shu*, 15.386.

19 Du Bao (fl. 7th century), edited and collated by Xin Deyong, *Daye zaji jijiao* (Xi'an: sanqin chubanshe, 2006), 49.

Landscapes for Travelling, Sightseeing, Wandering or Dwelling

1 Minna Törmä, *Enchanted by Lohans: Osvald Sirén's Journey into Chinese Art* (Hong Kong: Hong Kong University Press, 2013).

2 Osvald Sirén, *Kinas trädgårdar och vad de betytt för 1700-talets Europa*, 2 Vols. (Stockholm: Svenska Litteratur, 1948). The volumes in Swedish were published in English as *Gardens of China* (New York: Ronald Press, 1949) and *China and the Gardens of Europe of the Eighteenth Century* (New York: Ronald Press, 1950).

3 Ping Foong, *The Efficacious Landscape: On the Authorities of Painting at the Northern Song Court* (Cambridge, MA and London, England: Harvard University Asia Center, 2015).

4 Guo Xi, *Linquan gaozhi*, in *Zhongguo hualun leibian* (Chinese Painting Theory by Categories), vol. 1, ed. Yu Jianhua (Beijing: Renmin meishu chubanshe, 1957), 632–633; *Early Chinese Texts on Painting*, comp. and ed. Susan Bush and Hsio-yen Shih (Cambridge, MA and London: Harvard University Press, 1985), passages of the Guo Xi text are translated in chapter 4 "The Landscape Texts," 141–190.

5 Zong Bing is discussed in Bush and Shih, *Early Chinese Texts*, 36–38.

6 Guo Xi, *Linquan gaozhi*, 632–633; Bush and Shih, *Early Chinese Texts*, 151–152.

7 Published in Ju Hsi Chou, *Silent Poetry: Chinese Painting from the Collection of the Cleveland Museum of Art* (Cleveland: Cleveland Museum of Art, 2015), 34–43.

8 I am calling the internal viewer a Traveller with a capital "T" in order to distinguish him from the occasional travellers whom one meets along the way.

9 Attrib. to Yan Wengui, *Pavilions and Mansions by Rivers and Mountains (Jiangshan louguan tu)*, handscroll, ink and slight colour on paper, 32 × 161 cm, Osaka Municipal Museum of Art. For my analysis, see Minna Törmä, *Landscape Experience as Visual Narrative: Northern Song Landscape Handsrolls in the Li Cheng—Yan Wengui Tradition* (Helsinki: Academia Scientiarum Fennica, 2002).

10 Translated by A. C. Graham in *Two Chinese Philosophers*, new ed. (La Salle, IL: Open Court, 1992), 15.

11 I am using the original Swedish version as a source. Sirén, *Kinas trädgårdar*, Vol. I, 10. This analogy has become commonplace in writings on Chinese garden art as has the connection of garden design and painting in general. For a critique of this approach,

see Jing Xiao and Charlie Q. L. Xue. "Architecture in Ji Cheng's *The Craft of Gardens*: a visual study of the role of representation in counteracting the influence of the pictorial idea in Chinese scholar gardens of the Ming period," *Studies in the History of Gardens & Designed Landscape* 35, no. 3 (2015): 218–234.

12 Sirén, *Kinas trädgårdar*, Vol. I, 103.

13 My photos of Wangshiyuan accompanying this essay were taken in June 2009. However, the viewpoints were not chosen with this kind of sequential description in mind and I have added a number of details taken at the same time in Liuyuan.

14 The names of the locations are based on Alfreda Murck and Wen Fong, *A Chinese Garden Court: The Astor Court at the Metropolitan Museum of Art* (New York: Metropolitan Museum of Art, 1985), 11, 14.

15 Handscroll, ink and colour on silk, 53.9 × 127.8 cm; Palace Museum, Beijing: see http://en.dpm.org.cn/collections/collections/2013-01-24/1207.html (accessed 21.10.2018)

16 Aspects of this phenomenon have been explored by Wu Hung in *The Double Screen* (London: Reaktion Books, 1996).

17 Handscroll ink on silk, 29.5 × 71.1 cm; Nelson-Atkins Museum of Art, Kansas City: see https://art.nelson–atkins.org/objects/10336/whiling-away-the-summer-hsiaohsia-u?ctx=00292d92-75aa-4034-9884-8d990f29808b&idx=0 (accessed 21.10.2018). *The Double Screen* exists in a number of versions, but my reference is to the one which was formerly attributed to Zhou Wenju (active mid-10th century), but of Ming Dynasty in the 14th century; mounted as a handscroll, ink and colour on silk, 31.3 × 50 cm; Freer | Sackler (Gift of Charles Lang Freer) https://www.freersackler.si.edu/ob ject/F1911.195/ (accessed 21.10.2018).

18 The garden court (Astor Court) in the Metropolitan Museum of Art was designed after one of the courtyards in Wangshiyuan (see Figures 6–7) and the publication details the history of Wangshiyuan and its structure, as well as the construction of the Astor Court.

19 Chen Congzhou, *On Chinese Gardens* (New York: Better Link Press 2008), 15.

20 Murck and Fong, *Chinese Garden Court,* 30. Emphases in *italics* are mine.

21 Jing Xiao and Charlie Q. L. Xue, "Architecture in Ji Cheng's," 218, note that "pictorial aesthetic appears to suppress the corporeal operation of the eye to some extent."

22 The Art Institute in Chicago has one painting from this series in its collection: https://www.artic.edu/artworks/79314/dream-journey-to-rivers-and-mountains (accessed 21.11.2018).

23 Li Wai-yee."Gardens and Illusions from Late Ming to Early Qing,"*Harvard Journal of Asiatic Studies* 72, no. 2 (December 2012): 295–336.

24 Ibid., 298–299.

25 Ibid., 298–302: "Non-existent Garden" was "created" by Liu Shilong (17th century) and "Imagined Garden" by Dai Mingshi (1653–1713).

26 Ink and colour on paper, 116.6 × 23.8 cm; Art Institute of Chicago; see: https://www.artic.edu/artworks/40558/tea-drinking-under-the-wutong-tree?q=tang+yin (ac-

cessed 27.11.2018). It can also be viewed on https://scrolls.uchicago.edu/view-scroll/184 (accessed 27.11.2018), where you can digitally scroll the image as a handscroll.

Jingye: Crafting Scenes for the Chinese Garden

1 Although etymologically the word spectacle only refers to sight, I will use it here in the wider meaning for something which appeals to all the senses of the "spectator" or "viewer."

2 On the interaction between house and *garden, see Antoine Gournay, La Maison chinoise* (Paris: Klincksieck, 2016).

3 For example, see Edwin T. Morris, *The Gardens of China: History, Art and Meanings* (New York: Charles Scribner's Sons, 1983). Yang Hongxun, *Jiangnan yuanlin lun: Zhongguo gudian zaoyuan yishu yanjiu* (A treatise on the garden of Jiangnan: a study into the classical art of landscape design of China) (Shanghai: Shanghai renmin chuban she, 1994).

4 For a discussion on the many possible meanings of this term, see the entry for *jing* by Duncan Campbell in the "China Heritage Glossary," *China Heritage Quarterly*, no. 29 (March 2012), on line: http://www.chinaheritagequarterly.org/glossary.php?search term=029_jing.inc&issue=029

5 See Pauline Yu. *The Poetry of Wang Wei: New Translations and Commentary* (Bloomington: Indiana University Press, 1980). Patrick Carré, *Les saisons bleues: L' œuvre de Wang Wei poète et peintre* (Paris: Phébus, 1989).

6 William Chambers, *A dissertation on Oriental Gardening* (London: Griffin, 1772), 22.

7 Antoine Gournay, "Chine: jardins du Lingnan à la fin de la dynastie des Qing (1644–1911)," *Polia, Revue de l'art des jardins*, n° 1 (printemps 2004): 63–78.

8 On all these aspects, see Antoine Gournay, "Pour une analyse médiationniste du jardin chinois," *Tétralogiques*, 2018, n ° 23. *Le modèle médiationniste de la technique*. URL : http://tetralogiques.fr/spip.php?article93

9 Beijingshi yuanlin lühuaju, ed., *Beijing gushu mingmu* (Beijing: Yanshan chubanshe, 2009), 223.

10 Terese Tse Bartholomew, *Hidden meanings in Chinese Art* (San Francisco: Asian Art Museum of San Francisco, 2006), sections 7.41, 7.47–8.

11 Spectation will be used here to distinguish the devices used from those of the spectacle itself. I borrow the concept from Ph. Bruneau, "Qu'est-ce qu'une église?" *Ramage-Revue d'archéologie moderne et d'archéologie générale*, n° 9 (1991): 49–70. See its application to the Chinese garden in Antoine Gournay, "Le jardin chinois," *Ramage*, n° 12 (1994–1995): 119–135.

12 Jean-Denis Attiret. "Lettre à M. d'Assaut, 1er novembre 1743," in *Lettres édifiantes et curieuses écrites des missions étrangères*, vol. 28 (Paris, 1747), 1–49. English translation by Sir Harry Beaumont, *A Particular Account of the Emperor of China's Gardens Near Pekin* (London: Dodsley, 1752), 8. https://archive.org/details/aparticularacco00attigoog/page/n23

13 Chambers. *Dissertation on Oriental Gardening*, 64.

14 Ji Cheng, *The Craft of Gardens*, trans. Alison Hardie (New Haven and London: Yale University Press, 1988), 43.

15 On the art of *penjing* and its links with Taoist thought, see Rolf A. Stein, "Jardins en miniature d'Extrême-Orient," *Bulletin de* l'*École Française d'Extrême-Orient*, n° 42 (1943): 1–104; updated in *Le Monde en petit: Jardins en miniature et habitations dans la pensée religieuse d'Extrême-Orient* (Paris: Flammarion, 1987). English translation: *The World in Miniature: Container Gardens and Dwellings in Far Eastern Religious Thought*, trans. Phyllis Brooks (Stanford: Stanford University Press, 1990).

16 On the use of this term in the *Yuanye*, see and Wybe Kuitert, "Borrowing scenery and the landscape that lends—the final chapter of *Yuanye*," *Journal of Landscape Architecture* 10, no. 2 (2015): 32–43, on line: https://www.researchgate.net/publication/279245717_Borrowing_scenery_and_the_landscape_that_lends_-_The_final_chapter_of_Yuanye

17 Hu Dongchu, *The Way of the Virtuous: The Influence of Art and Philosophy on Chinese Garden Design* (Beijing: New World Press, 1991), 93–95. For a more detailed discussion on the role of these openings, see Antoine Gournay, "Le système des ouvertures dans l'aménagement spatial du jardin chinois," in *Extrême-Orient, Extrême-Occident*, n° 22, *L'art des jardins dans les pays sinisés, Chine, Japon, Corée, Vietnam*, ed. I Léon Vandermeersch (2000): 51–71.

18 English translation by Hardie in Ji Cheng, *Craft of Gardens*, 76.

19 Wang Yi. *Yuanlin yu Zhongguo wenhua* (Shanghai: Shanghai renmin chubanshe, 1990). Translated by Yu Luo Rioux as *A Cultural History of Classical Chinese Gardens* (New York: SCPG Publishing Corporation, 2015), figures 1–8 show a scene with the same title in Qiuxia pu, in Shanghai.

"Knowing the Painting Vein" and "Being Like in a Ravine": Jichangyuan Landscape Reconstruction in Early Qing and "Zhang's Mountain" in 17th Century Jiangnan

1 The original version of this essay was published as Gu Kai, "'Knowing the Painting Vein' and 'Being Like in a Ravine': Jichangyuan Landscape Reconstruction in Early Qing and 'Zhang's Mountain' in 17th Century Jiangnan," *Zhongguo yuanlin* 35 (July 2019): 124–129, with minor modifications.

2 Zhu Youjie (1919–2015) said, "I admire Jichangyuan the most among all the famous remaining Ming and Qing gardens in Jiangnan," in "Foreword" to *Xishan Qinshi Jichangyuan wenxian ziliao changbian*, ed. Qin Zhihao (Shanghai: Shanghai cishu chubanshe, 2009). In all the seven editions of *History of Chinese Architecture* edited by Pan Guxi, still the popular textbook in Chinese universities, Jichangyuan is always the first case in the part on "Private Gardens in Ming and Qing Jiangnan." Cao Xun selected "five top gardens in China" with Jichangyuan as the first one, in "Jiangnan yuanlin jia tianxia, Jichangyuanlin jia Jiangnan," *Fengjing yuanlin* 25 (November 2018): 14. Dong Yugan takes "Jichangyuan in Wuxi as the top one among all gardens I have visited" in "Shuangyuan bafa: Jichangyuan yu Xiequyuan bidui," *Jianzhu shi* 172 (December 2014): 108; etc.

3 See Huang Xiao, "Xishan Qinshi Jichangyuan kao" (Master Thesis, Beijing University, 2010), 96.

4 Huang Xiao and Liu Shanshan, "Mingdai houqi Qinyao Jichangyuan lishi yange kao," in *Jianzhu shi* di 28 ji, ed. Jia Jun (Beijing: Qinghua daxue chubanshe, 2012), 112–135.

5 Xu Yongyan, Qin Songlin and Yan Shengsun, eds., *Wuxi xian zhi* (1690), 7.9b.

6 Huang Xiao and Liu Shanshan, "Jichangyuan: yizuo yuanlin beihou de wenming lunhui," *Zhonghua yichan* 104 (June 2014): 46–65.

7 Qin Zhihao, "Qingchu Jichangyuan de guibing he gaizhu," in *Jichangyuan jianyuan 490 zhounian yantaohui lunwenji* (Wuxi, October 2017), 1–9.

8 Xu, Qin and Yan, *Wuxi xian zhi*, 7.9b.

9 Huang Xiao, Liu Shanshan, and Qin Shaoying, "Zhang Nanyuan de shengping, jiazu yu jiaoyou: qiaolai *Zhang Chushi muzhiming* kaolue," *Fengjing yuanlin* 27, no.4 (April 2020): 118–125.

10 Cao Xun, "Zhang Nanyuan de zaoyuan dieshan zuopin," in *Zhongguo jianzhu shi lun huikan*, di er ji, ed. Wang Guixiang (Beijing: Qinghua daxue chubanshe, 2009), 327–378.

11 Gu Kai, "Chongxin renshi Jiangnan yuanlin: zaoqi chayi yu wanming zhuanzhe," *Jianzhu xuebao* (S1, 2009): 106–110.

12 Cao, "Zhang Nanyuan de zaoyuan dieshan zuopin," 327.

13 Cao Xun, "Zaoyuan dashi Zhang Nanyuan: jinian Zhang Nanyuan dansheng sibai zhounian (1)," *Zhongguo yuanlin* (no. 1, 1988): 21–26.

14 Wu, *Qishier feng zuzheng ji (Wenji)*, 1745 ed., 10.14a.

15 Cao Xun, "Zaoyuan dashi Zhang Nanyuan: jinian Zhang Nanyuan dansheng sibai zhounian (2)," *Zhongguo yuanlin* (no. 3, 1988): 2–9.

16 Gu Kai, "Huayi yuanze de queli yu wanming zaoyuan de zhuanzhe," *Jianzhu xue bao* (S1, 2010): 127–129.

17 Wu Weiye, *Wu Meicun quanji*, ed. Li Xueying (Shanghai: Shanghai guji chubanshe, 1990), 1059–61.

18 Wai-Yee Li, "Gardens and Illusions from Late Ming to Early Qing," *Harvard Journal of Asiatic Studies* 72, no.2 (December 2012): 319.

19 Cao, "Zaoyuan dashi Zhang Nanyuan (2)," 5–6.

20 Wen C. Fong, *Beyond Representation: Chinese Painting and Calligraphy 8th–14th Century* (New York: Metropolitan Museum, 1992), 441–97.

21 Gu Kai, "Niru huazhong xing: wanming jiangnan zaoyuan dui shanshui youguan tiyan de kongjian jingying yu huayi zhuiqiu," *Xin jianzhu* (no. 6, 2016): 44–47.

22 Dai Mingshi, *Nanshan ji,* in *Xuxiu siku quan jibu* 1419 (Shanghai: Shanghai guji chubanshe, 2002), 143.

23 Cao, "Zaoyuan dashi Zhang Nanyuan (2)," 4.

24 Ji Cheng, *Craft of Gardens*, Trans. Alison Hardie (Shanghai Press and Publishing

Development Company, 2012), 118.

25 Gu, "Niru huazhong xing," 44–47.

26 Huang Zongxi, *Huang Zongxi quanji, vol.10, Nanlei shiwen ji shang*, ed. Shen Shanhong (Hangzhou: Zhejiang guji chubanshe, 2012), 571.

27 Shao Changheng, *Shao Zixiang quanji,* in *Siku quanshu cunmu congshu jibu* 247 (Jinan: Qilu shushe, 1997), 715.

28 Qin Zhihao ed., *Xishan Qinshi Jichangyuan wenxian ziliao changbian* (Shanghai: Shanghai cishu chubanshe, 2009), 82.

29 Gu Kai, *Mingdai Jiangnan yuanlin yanjiu* (Nanjing: Dongnan daxue chubanshe, 2010), 183–84.

30 The war, as a result of which the Qing Dynasty was established, caused great damage to the Jiangnan society and economy. See Willard J. Peterson ed., *The Cambridge History of China, Vol. 9: The Ch'ing Dynasty, Part 1: To 1800* (Cambridge: Cambridge University Press, 2002), 86–89.

31 There are two judgments on the time of the reconstruction in the early Qing:"between 1667 and 1668" by Huang Xiao (see Huang Xiao, *Xishan Qinshi Jichangyuan kao*, 96) and "between 1666 and 1667" by Qin Zhihao (see Qin Zhihao, "Qingchu Jichangyuan de guibing he gaizhu," 7). Both acknowledge that it lasted less than two years.

32 Liang Jie and Zheng Xin, "Wanming Jichangyuan shuichi 'Jinhuiyi' jiqi zhoubian fuyuan yanjiu," *Zhongguo yuanlin* 34 (December 2018): 138.

33 Shao, *Shao Zixiang quanji*, 715.

34 This pavilion name in the previous layout was kept but used in a different building after the reconstruction.

35 Qin, *Xishan Qinshi Jichangyuan*, 126.

36 Ibid., 131.

37 Ibid., 129.

38 James Cahill, *Hills Beyond a River: Chinese Painting of the Yuan Dynasty, 1279–1368* (New York & Tokyo: John Weatherhill Inc., 1976), 117–18.

39 Shao Changheng, *Shao Zixiang quanji*, 715. Zijiu is the courtesy name of Huang Gongwang.

40 The prominent scene of a slanting tree can be seen in various paintings in the Qing period and is still well kept today.

41 The bridge was called "Zigzag Bridge" in one of Wu Qi's poems in the early Qing (see Huang Xiao, *Xishan Qinshi Jichangyuan kao*, 123). In various paintings of the middle Qing period, the bridge has three turns. So the present straight bridge is not in the original form. And the present covered bridge at the northeast corner of the pond does not appear in Qing images.

42 Qin, *Xishan Qinshi Jichangyuan*, 88.

43 Ibid., 10.

44 Ibid., 24.

45 Ibid., 35.

46 Ibid., 33.

47 Ibid., 34.

48 Ibid., 42.

49 Xu Zuanzeng, *Baolun tang gao,* in *Xuxiu siku quanshu* jibu 1409 (Shanghai: Shang hai guji chubanshe, 2002), 651.

50 Qin, Xishan Qinshi Jichangyuan, 81.

51 Ibid., 92.

52 Qin, "Qingchu Jichangyuan," 6.

53 Gu Kai, "Mingmo Qingchu Taicang Lejiaoyuan shiyi pingmian fuyuan tanxi," *Fengjing yuanlin* 139 (February 2017): 25–33.

54 Xu, *Wuxi xian zhi* , 7.9b.

55 Li Nianci, *Gukou shanfang shiji, in Siku quanshu cunmu congshu, jibu* 232 (Jinan: Qilu shushe, 1997), 794.

56 The scene is called "Gully of Eight Tones" today, which is often regarded as a name since it was created. But this name did not appear until the 20th century according to the remaining historical materials.

57 Huang, *Xishan Qinshi Jichangyuan kao*, 118.

58 Qin, *Xishan Qinshi Jichangyuan*, 123.

59 Qin, "Qingchu Jichangyuan," 6.

60 Qin, *Xishan Qinshi Jichangyuan*, 84.

61 Chen Congzhou, *On Gardens* (Shanghai: Tongji University Press, 1984), 40.

Gardens Real and Imagined: Gardens, Landscape Painting, and Biehaotu

1 On "gifts" in Chinese art at this period, see Craig Clunas, *Elegant Debts: The Social Art of Wen Zhengming* (London: Reaktion Books, 2004).

2 Zhang Chou, *Qinghe shuhua fang*, written in 1616, *Qinding siku quanshu*, 11.42b, available at https://ctext.org/library.pl?if=gb&res=6240 (accessed 29 October 2018). Emphasis is mine.

3 On garden culture in the Ming, see Craig Clunas, *Fruitful Sites: Garden Culture in Ming Dynasty China* (London: Reaktion Books, 1996).

4 Wen Zhengming, *Wen Zhengming ji*, ed. Zhou Daozhen (Shanghai: Shanghai guji

chubanshe, 1987), 502, quoted in Clunas, *Elegant Debts,* 99.

5 Liu Jiu'an, "Wumen huajia zhi biehaotu ji jianbie juli," *Gugong Bowuyuan yuankan* 49 (issue 3, 1990): 54–61.

6 Xu Ke, "Junzi yu shi: Wen Zhengming biehaotu zhong de tujie yu yinyu," *Meishu xuebao* (issue 5, 2016): 33–40; ibid., "Wumen huapai biehaotu de wenhua yiyi yu shehui gongyong—yi Wen Zhengming biehaotu wei li," *Nanjing yishu xueyuan xuebao (meishu yu sheji)* (issue 6, 2016): 55–61.

7 http://www.dpm.org.cn/collection/paint/228141.html?hl=%E5%8F%8B%E6%9D%BE%E5%9C%96 (accessed on 12 October 2018).

8 Zhang Chou, *Qinghe shuhua fang, 12shang*.7b.

9 An image of the complete scroll is available at: https://m–minghuaji.dpm.org.cn/paint/appreciate?id=08f185a3d4e3467cb073832559f59447 (accessed 12 October 2020).

10 Wen Zhengming, *Wen Zhengming ji*, 582, quoted in Clunas, *Elegant Debts*, 44.

11 Richard Edwards, *The Art of Wen Cheng-ming (1470–1559)* (Ann Arbor, MI: University of Michigan Museum of Art, 1976), 138–139.

12 An image of the complete scroll is available at: https://m–minghuaji.dpm.org.cn/paint/appreciate?id=41ebf4498dfa4d58b89705e4d8dcf8a4 (accessed 12 October 2020).

13 Liu Jiu'an, "Wumen huajia zhi biehaotu", 58.

14 Ji Cheng, ed. Chen Zhi, *Yuanye zhushi,* 2nd ed. (Beijing: Zhongguo jianzhu gongye chubanshe, 1988), 47.

15 Wu Weiye, *Wu Meicun quanji*, ed. Li Xueying (Shanghai: Shanghai guji chuban she, 1990), 1060. Translation from Alison Hardie, "The Life of a Seventeenth-Century Chinese Garden Designer: *The Biography of Zhang Nanyuan,* by Wu Weiye (1609–71)," *Garden History* 32, no. 1 (Spring 2004): 138–140.

16 I use "Man of the Way" rather than "Daoist" as the translation for *daoren*, since by this time Wu Li was in the process of converting to Catholic Christianity, a different "way" from that of traditional Chinese philosophy.

17 http://www.dpm.org.cn/collection/paint/228483.html?hl=%E6%B1%82%E5%BF%97%E5%9C%92 (accessed 12 October 2018).

18 https://insitu.arthistory.wisc.edu/Details/10462/269909 (accessed 12 October 2020). On Wen Zhengming's relationship with Zhang Fengyi and his younger brother Xianyi (d. 1604), see Clunas, *Elegant Debts*, 130.

19 Wang Shizhen, *Yanzhou shanren sibugao*, completed 1577 (Taipei: Weiwen tushu chubanshe, 1976), 3581. Full text and translation in Alison Hardie, "Chinese Garden Design in the Later Ming Dynasty and its Relation to Aesthetic Theory" (unpublished DPhil thesis, University of Sussex, 2001), vol.3 (Appendix), 105–109. A version of my translation also appears in Alison Hardie and Duncan M. Campbell eds., *The Dumbarton Oaks Anthology of Chinese Garden Literature* (Washington DC: Dumbarton Oaks Research Library and Collection, 2020), 361–363.

20 http://painting.npm.gov.tw/Painting_Page.aspx?dep=P&PaintingId=4016 (accessed 12 October 2018).

21 Wang Shizhen, "Yanshanyuan ji," in idem, *Yanzhou shanren xugao*, contained in *Mingren wenji congkan*, ed. Shen Yunlong (Taipei: Wenhai chubanshe, 1970), 2957–97. See also Chen Zhi and Zhang Gongchi, eds., *Zhongguo lidai ming yuan ji xuanzhu* (Hefei: Anhui kexue jishu chubanshe, 1983), 130–156. Full text and translation in Hardie, *Chinese Garden Design*, vol.3 (Appendix), 8–76: translation also in Hardie and Campbell, eds., *Dumbarton Oaks Anthology*, 321–346.

22 Kate Kerby and Mo Zung Chung, *An Old Chinese Garden: A Three-fold Masterpiece of Poetry, Calligraphy and Painting* (Shanghai: Chung Hwa Book Co., 1922). A fairly good online reproduction of the images can be found at http://blog.sina.com.cn/s/blog_647cbe2b0100h6kq.html (accessed 18 October 2018); the Sophora Rain Pavilion is near the end.

23 Wen Zhengming, *Wen Zhengming ji*, ed. Zhou Daozhen (Shanghai: Shanghai guji chubanshe, 1987), 1276.

24 Ibid., 1210.

25 https://www.shanghaimuseum.net/museum/frontend/collection/zoom.action?cpInfoId=949&picId=1863 (accessed 12 October 2018).

26 Wen Zhengming, *Wen Zhengming ji*, 1303.

27 http://www.nipic.com/show/2/27/4803652k08eec23e.html (accessed 12 October 2018).

28 Translation by Richard Edwards in idem et al., *Art of Wen Cheng-ming*, 150–151; this also has a colour reproduction. See also https://www.metmuseum.org/art/collection/search/39536 (accessed 12 October 2020).

29 No image is currently available on the Palace Museum website. There is a good colour plate (including Qian's colophon) in Liu Rushi, *Liu Rushi ji*, ed. Zhou Shutian and Fan Jingzhong (Hangzhou: Zhongguo meishu xueyuan chubanshe, 2002).

30 James Cahill, *The Distant Mountains: Chinese Painting of the Late Ming Dynasty, 1570–1644* (New York & Tokyo: Weatherhill, 1982), 132. No image is currently available on the Palace Museum website. A version of the painting can be seen at http://auc tion.artron.net/paimai-art00270113/ (accessed 23 October 2018).

31 Cahill, *Distant Mountains*, 39–40.

32 James Cahill, *The Compelling Image: Nature and Style in Seventeenth-Century Chinese Painting* (Cambridge, MA and London, England: Harvard University Press, 1982), 6–7.

33 http://www.dpm.org.cn/collection/paint/229436.html (accessed 18 October 2018).

34 http://www.dpm.org.cn/collection/paint/228964.html?hl=錢穀 (accessed 18 October 2018).

35 Cao Xun, "Qingdai zaoyuan dieshan," in Jianzhu lishi yu lilun di er ji, ed. Zhongguo jianzhu xuehui and Jianzhu lishi xueshu weiyuanhui (Nanjing: Jiangsu renmin chu banshe, 1981), 116.

36 The full album is reproduced in colour in June Li and James Cahill, *Paintings of Zhi Garden by Zhang Hong: Revisiting a Seventeenth-Century Chinese Garden* (Los Angeles: Los Angeles County Museum of Art, 1996); there is also a black and white

reproduction of the holistic view in Cahill, *Compelling Image*, 20. See also http://www.zhigarden.com/PicDetail.aspx?ID=181 (accessed 12 October 2020).

37 Cahill, *Compelling Image*, especially Chapter 3; the section on Xiang Shengmo is on 98–105.

38 Cahill, *Distant Mountains*, 86.

39 Zhao Xiaohua,"Shi you biehao ru hua lai—Tan Yin *Wuyangzi yangxing tu* kaobian," *Wenwu* (issue 8, 2000): 78–83.

40 Han Jinlei, "Tang Yin "biehaotu" chuangzuo zhong 'jing' yu 'ren' de zugou guanxi," *Cuiyuan* (*Minzu wenxue*) (issue 4, 2016): 76–78.

41 Ye Yu, "Biehaotu yu shanshuihua de xiangzheng yiyi—cong Wen Zhengming de '*Zhenshangzhai tu*' tanqi," *Xin meishu* (issue 12, 2016): 16–23.

Hills and Ravines in the Heart: A World without "Space"

1 Liu Dunzhen, "Suzhou de yuanlin," *Nanjing gongxueyuan yuankan* no. 4, special issue (1957): 8, 11, 12 and 15.

2 Liu Dunzhen, *Suzhou gudian yuanlin* (Beijing: Zhongguo jianzhu gongye chubanshe, 1979), 9–10. The volume was compiled and edited by Pan Guxi.

3 Tong Jun does not refrain from talking about space in his "Architecture Chronicle," *T'ien Hsia Monthly* V, no. 3 (October 1937): 308–312. I consulted the version included in *Tong Jun wenji* 1 (Beijing: Zhongguo jianzhu gongye chubanshe, 2000), 81–88.

4 Tung Chui, "Chinese Gardens: Especially in Kiangsu and Chekiang," *T'ien Hsia Monthly* III, no. 3 (October 1936): 220: "The Chinese garden is primarily not a single wide open space, but is divided into corridors and courts, in which buildings, and not plant life, dominate." In this context, Fang Yong and Wang Tan's translation of the word *kongjian*, in Tong Jun, *Yuan lun* (Tianjin: Baihua wenyi chubanshe, 2006), 1, is very accurate.

5 Tong Jun, *Jiangnan yuanlin zhi*, 2nd ed. (Beijing: Zhongguo jianzhu gongye chu banshe, 1987).

6 Tong Jun, "Suzhou Gardens: Which Embody All the Characteristics of the Chinese Landscape Art," in *Tong Jun wenji* 1 (Beijing: Zhongguo jianzhu gongye chu banshe, 2000), 262–274, quoted sentences on 269; Chinese version"Suzhou yuanlin: Ji Zhongguo zaoyuan yishu tezheng yu yiti," trans. Li Daxia and Wang Tan, in ibid., 275–282. Similarly, references to "space" appear in his "Glimpses of Gardens in Eastern China," known in Chinese as *Dongnan yuanshu*, completed in 1983 but published in 1997 (Beijing: Zhongguo jianzhu gongye chubanshe).

7 For a comparative reading of Tong Jun's and Liu Dunzhen's studies of Chinese gardens, see Gu Kai, "Tong Jun yu Liu Dunzhen de Zhongguo yuanlin yanjiu bijiao," *Jianzhushi* 173 (February 2015): 92–105.

8 "As a term," Adrian Forty reminds us in his *Words and Buildings: A Vocabulary of Modern Architecture* (London: Thames & Hudson, 2000), 256, "'space' simply did not exist in the architectural vocabulary until the 1890s."

9 Clifford Geertz, *The Interpretation of Cultures: Selected Essays* (New York: Basic Books, 1973), 5.

10 F.W. Mote, "The Arts and the 'Theorizing Mode' of the Civilization," in *Artists and Tradition: Uses of the Past in Chinese Culture*, ed. Christian F. Murck (Princeton, NJ: Princeton University Press, 1976), 5.

11 For instance, Minna Törmä, "Looking at Chinese Landscape Painting," in *Looking at Other Cultures: Works of Art as Icons of Memory*, ed. Anja Kervanto Nevanlinna (Helsinki: Taidehistorian Seura, 1999), 119–135; Andong Lu, "Lost in translation: Modernist Interpretation of the Chinese Garden as Experiential Space and Its Assumptions," *The Journal of Architecture* 16, no. 4 (2011): 499–527.

12 James Elkins, *The Poetics of Perspective* (Ithaca, NY: Cornell University Press, 1994), 41.

13 Erwin Panofsky, *Perspective as Symbolic Form*, trans. Christopher S. Wood (New York: Zone Book, 1991), 27. For a discussion in this context of the issue of homogeneous space, see Branko Mitrović, "Leon Battista Alberti and the Homogeneity of Space," *Journal of the Society of Architectural Historians* 63, no. 4 (December 2004), 424–425.

14 James Elkins, "Renaissance Perspectives," *Journal of the History of Ideas* 53, no. 2 (April–June 1992): 209, 217.

15 William V. Dunning, *Changing Images of Pictorial Space: A History of Spatial Illusion in Painting* (New York: Syracuse University Press, 1991), 65.

16 Marco Musillo, *The Shining Inheritance: Italian Painters at the Qing Court, 1699–1812* (Los Angeles, CA: The Getty Research Institute, 2016), 120–121, 126, dismisses as "oversimplified" another account of the opposition between the European and Chinese pictorial experiences, namely the formers̓ use of perspective being considered a technique that elicited physical reality, in contrast to the latter's tradition of searching for spiritualpoetic aspects of painting. Musillo demonstrates that searching for spirituality in European painting is no less evident in the early modern artistic literature, even though I find odd his universalistic assertion made in the form of a rhetorical question "Mimesis and Poiesis: is not the interaction between these two forms of human expression an issue common to the Chinese and the European painting traditions?" (127) One should also note that, on the one hand, the early development of European perspective drawing was associated with the idea of geometrically representing an ideal Christian world in order to reveal the orderly and logical principles on which God created the world, and of creating what Edgerton calls "pictures in the service of God"; and that, on the other hand, linear perspective was first devised with no clear scientific application in mind, but solely to help solve a very medieval theological problem—to remove the old iconic barrier between the world of the viewers and the realm of the image, so that God and his saints were once more immanent in the viewers' daily lives, and could be seen and touched as if they were actual life-size persons in the here and now. Cf., e.g., Samuel Edgerton, *The Renaissance Rediscovery of Linear Perspective* (New York: Basic Books, 1975), 16, 157; Dunning, *Changing Images*, 58; D.E. Mungello, *The Great Encounter of China and the West*, 1500–1800, 4th ed. (Plymouth, England: Rowman & Littlefield, 2013), 40; Kristina Kleutghen, *Imperial Illusions: Crossing Pictorial Boundaries in the Qing Palaces* (Seattle, WA: The University of Washington Press, 2015), 80.

17 For these debates, cf. Panofsky, *Perspective as Symbolic Form*; Edgerton, *Renaissance Rediscovery*; Martin Kemp, *The Science of Art: Optical Themes in Western Art from Brunelleschi to Seurat* (New Haven, CT: Yale University Press, 1990), esp. 7–162; Hubert Damisch, *The Origin of Perspective*, trans. John Goodman (Cambridge, MA: The MIT Press, 1994); James Elkins, "Renaissance Perspectives" and *The Poetics of Perspective*;

Mitrović, "Leon Battista Alberti", 424–439.

18 Norman Bryson, *Word and Image: French Painting of the Ancien Régime* (Cambridge, England: Cambridge University Press, 1981), 89–91. This view is fully taken by Dunning, *Changing Images*, 104.

19 Joseph Needham, *Science and Civilisation in China* Volume IV, Part 3 (Cambridge, England: Cambridge University Press, 1971), 114–115.

20 James O. Caswell,"Lines of Communication: Some' Secrets of the Trade' in Chinese Painters' Use of 'Perspective'," *RES* 40 (Autumn 2001): 189, 190.

21 For a survey of recent scholarship on the art and visual culture of the High Qing Dynasty in the context of Sino-European artistic interactions, see Cheng-Hua Wang, "Global Perspective on Eighteenth-Century Chinese Art and Visual Culture," *The Art Bulletin* 96, no. 4 (2014): 379–394. For a brief discussion of the conditions for and development of linear perspective in Chinese art during the eighteenth century, see Kristina Kleutghen, "From Science to Art: The Evolution of Linear Perspective in Eighteenth-Century Chinese Art," in *Qing Encounters: Artistic Exchanges between China and the West*, ed. Petra Ten-Doesschate Chu and Ning Ding (Los Angeles, CA: The Getty Research Institute, 2015), 173–189.

22 For brief introductions to Masanobu's historical achievements, see Hugo Munsterberg, *The Japanese Print: A Historical Guide* (New York: Weatherhill, 1982), 37–45; Andreas Marks, *Japanese Woodblock Prints: Artists, Publishers and masterworks 1680–1900* (North Clarendon, VT: Tuttle Publishing, 2010), 32–35.

23 Timon Screech, "The Meaning of Western Perspective in Edo Popular Culture," *Archives of Asian Art* 47 (1994): 59. Screech, again referring to the same picture, puts his conjecture in another way in his *The Lens within the Heart: The Western Scientific Gaze and Popular Imagery in Later Edo Japan* (Honolulu, HI: University of Hawai'i Press, 2002), 104: "His earliest works are not of foreign vistas, but show Japanese places and tend to be interior scenes, probably through self-restrictions imposed by an artist not yet confident enough to render the more flexed and uneven lines of the irregular outdoors."

24 Stephen Little, "The Lure of the West: European Elements in the Art of the Floating World," *Art Institute of Chicago Museum Studies* 22, no. 1 (1996): 78–80. Harold P. Stern, however, defers any value judgment of this kind when succinctly analysing another work of Masanobu's, in his "A Private Puppet Performance," in *Freer Gallery of Art Fiftieth Anniversary Exhibition I. Ukiyo-e Painting* (Washington D.C.: Smithsonian Institution, 1973), 91.

25 Screech, *Lens within the Heart*, 104.

26 For an introduction to and translation of the text on Shuten-dōji, see Noriko T. Reider, "Shuten Dōji: 'Drunken Demon'," *Asian Folklore Studies* 64, no. 2 (2005): 207–231.

27 Yoriko Kobayashi-Sato and Mia M. Mochizuki, "Perspective and Its Discontents or St. Lucy's Eyes,"in *Seeing across Cultures in the Early Modern World*, ed. Dana Leibsohn and Jeanette Favrot Peterson (Surrey, England: Ashgate, 2012), 28.

28 Ibid., 35.

29 Angus Lockyer, "Hokusai's Thought," in *Hokusai: Beyond the Great Wave*, ed.

Timothy Clark (London, England: Thames & Hudson, 2017), 33.

30 Screech, "Meaning of Western Perspective," 66–67.

31 Kobayashi-Sato and Mochizuki, "Perspective and Its Discontents," 32, 35–36. The authors cite the title of the print as *The First Night Kyōgen at the Nakamura Theatre* 中村座顔見世狂言.

32 Ibid., 36–37.

33 Ibid., 39, Figure 1.8.

34 Katsushika Hokusai, *Hokusai Manga* (Nagoya: Eirakuya Toshiro, 1815), 3.15b–16a. The British Museum free image service, 1979,0305,0.428.3. David Bell, *Ukiyo-e Explained* (Kent, England: Global Oriental, 2004), 210, regards this technique as "something of a curiosity," while Kobayashi-Sato and Mochizuki, "Perspective and Its Discontents," 38–42, have implied that vanishing-axis perspective was an innovation on the part of the ukiyo-e artists, and, to add force to their argument, have cited as an example the frontispiece of Sugita Genpaku's *A Bew Book of Anatomy (Kaitai shinsho)* (1774), which "copied quite literally by Odano Naotake from a Netherlandish edition of Juan de Valverde's 1556 and 1560 *La anatomia del corporis humani* ..." This, however, does not seem to have been the case; that is, the fishbone perspective technique may neither be a curious approach of the Ukiyo-e artists nor an invention by Hokusai. Many a book on human anatomy published in Europe in the mid–16th century, such as Juan Valverde de Amusco's *Anatomie, oft levende beelden vande deelen* ... (1568), contains a frontispiece whose image clearly displays the employment of vanishing-axis perspective. What can be certain is that the ukiyo-e artists happily adopted this approach, as it worked well for them.

35 Kobayashi-Sato and Mochizuki, "Perspective and Its Discontents," 3.

36 Anita Chung, *Drawing Boundaries: Architectural Images in Qing China* (Hawai'i, HI: University of Hawai'i Press, 2004), 94.

37 This statement could be misleading, as if the format was the cause for the mode of perception. I would argue, although the limited space in this essay does not allow me to do so in detail, that the particular format and the distinctive mode of perception are both conditions for and results of each other.

38 Musillo, *The Shining Inheritance: Italian Painters*, 89–92.

39 Maxwell K. Hearn, "Art Creates History: Wang Hui and *The Kangxi Emperor's Southern Inspection Tour,*" in *Landscapes Clear and Radiant: The Art of Wang Hui* (1632–1717), ed. Maxwell K. Hearn (New York: The Metropolitan Museum of Art, 2008), 180–181.

40 Needham, *Science and Civilisation*, 112.

41 But Needham (ibid., 381), commenting on Muslim influence on Chinese astronomy in the Yuan Dynasty, does indicate that"The Chinese did not adopt the advanced geometry needed for the stereometric projections on astrolabes and sundials," and also states that "the Jesuits brought a clear exposition of the geometrical analysis of planetary motions, and of course the Euclidean geometry necessary for applying it" (ibid., 437).

42 Neither pre-modern Chinese dictionaries (such as *Peiwen yunfu* and *Jingji zuangu*) nor modern Chinese dictionaries dedicated to classical Chinese (such as *Ciyuan*) have

an entry for *kongjian*. The word is contained in both *Cihai* and *Hanyu da cidian*, but its definitions and etymology are entirely modern.

43 Dōshin Satō, *Modern Japanese Art and the Meiji State: The Politics of Beauty*, trans. Hiroshi Nara (Los Angeles, CA: The Getty Research Institute, 2011), 271.

44 Xu Shen, *Shuowen jiezi*, completed in 100–121 CE (Beijing: Zhonghua shuju, 1963), 12A.248. It is not difficult for one to share the view of Cordell D.K. Yee, "Chinese Cartography among the Arts: Objectivity, Subjectivity, Representation," in *The History of Cartography* Vol. 2, Book 2, ed. J. B. Harley and David Woodward (Chicago, IL: Chicago University Press, 1994), 139, that, in the context of his discussion on the ideograph *hua* (picture, painting), "The *Shuowen* may not be a reliable guide to etymology, but it at least serves as a useful indicator of Han opinion and misconceptions about it."

45 Duan Yucai (1735–1815), *Shuowen jiezi zhu*, completed in 1808 and published in 1815 (Shanghai: Shanghai guji chubanshe, 1981), 12A.589 indicates that in order to differentiate the two, the ideograph is later pronounced *xian* when it is used for the second meaning. Infrequent occurrences in classical documents, such as *Guanzi* and *Lüshi chunqiu*, of combining the two characters into a word all refer to "unoccupied" or "in leisure."

46 Translation by Burton Watson, *The Complete Works of Chuang Tzu* (New York: Columbia University Press, 1968), 256. For the text in Chinese, see Wang Xianqian (1842–1917), *Zhuangzi ji jie* (Beijing: Zhonghua shuju, 1987), 23.203.

47 John S. Major, et al., trans. and ed., *The Huainanzi: A Guide to the Theory and Practice of Government in Early Han China*, by Liu An, *King of Huainan* (New York: Columbia University Press, 2009), 415. In Chapter 3 of the same book, Major translates *yuzhou* simply as "space-time" (114 and ibid., note 1), which, instead of "captur[ing] the idea very well," seems even more misleading. For the quoted text in Chinese, see He Ning, *Huainanzi ji shi* (Beijing: Zhonghua shuju, 1998), 11.798. Angus C. Graham's understanding of the characters in this context, "process enduring in time" and "matter extending in space," is much more sensible, as the terms "time" and "space" here merely take the allusive function of suggesting that the duration and extension happen in what we now loosely call time and space. See A.C. Graham, "Reflections and Replies," in *Chinese Texts and Philosophical Contexts: Essays Dedicated to Angus C Graham*, ed. Henry Rosemont Jr. (La Salle, IL: Open Court, 1990), 279.

48 He Ning, *Huainanzi ji shi*, 6.469. Major (*The Huainanzi*, 221) translates the sentence in this way: "So the swallows and sparrows mocked them, saying that they were incapable of matching them in squabbling among the roof beams and rafters." Major then states (ibid., note 47): "The text here makes a play on words: *yuzhou* 宇宙 is used literally in its sense of 'roof beams and rafters' (e.g., of a barn, the natural habitat of swallows and sparrows), but it also calls to mind its more usual meaning of 'universe' (i.e., the habitat of phoenixes)." The identification of the play on words is insightful indeed. On the other hand, one can hardly regard *yuzhou*'s meaning of "universe" as "more usual"—the "literal" meanings of *yu* and *zhou* must have been more usual; otherwise there would have been no need for Gao You 高誘 (fl. early 3rd century CE) and other later commentators to explain the derived meanings of the ideographs in those contexts where they refer to "universe."

49 Richard E. Strassberg, *Inscribed Landscapes: Travel Writing from Imperial China* (Los Angeles, CA: University of California Press, 1994), 138, 143, 163. The emphases are mine. Cf. Ouyang Xiu, *Ouyang Xiu quanji* (Beijing: Zhonghua shuju, 2001), 39.576–577; Bai Juyi, Bai Juyi ji (Beijing: Zhonghua shuju, 1979), 43.940–941; Liu Zongyuan, *Liu Zongyuan ji* (Beijing: Zhonghua shuju, 1979), 29.765–766.

50 James Cahill, *Hills Beyond a River: Chinese Painting of the Yüan Dynasty, 1279–1368* (New York: Weatherhill, 1976), 16. The emphasis is mine.

51 Guo Ruoxu, *Tuhua jianwen zhi,* in *Huashi congshu*, ed. Yu Anlan (Shanghai: Shanghai renmin meishu chubanshe, 1962), 1.1.6.

52 Alexander Coburn Soper, *Kuo Jo-Hsü's Experiences in Painting: An Eleventh Century History of Chinese Painting Together with the Chinese Text in Facsimile* (Washington, D.C.: American Council of Learned Societies, 1951), 12. The emphasis is mine.

53 Susan Bush and Hsio-yen Shih, *Early Chinese Texts on Painting* (Hong Kong: Hong Kong University Press, 2012), 111. The emphasis is mine.

54 Robert J. Maeda, "Chieh–hua: Ruled–line Painting in China," *Ars Orientalis* 10 (1975): 123. The emphasis is mine.

55 Caswell, "Lines of Communication," 201. The emphasis is mine.

56 Duncan M. Campbell, "Zheng Yuanxun's 'A Personal Record of My Garden of Reflections'," *Studies in the History of Gardens & Designed Landscapes* 29, no. 4 (2009): 280, note 18. The emphasis is mine. The original text can be found in Li Dou, *Yangzhou huafang lu*, punc. and proofreading Wang Beiping and Tu Yugong (Beijing: Zhonghua shuju, 1960), 6.145; and Li Dou, *Yangzhou huafang lu* (Yangzhou: Jiangsu Guangling guji keyinshe, 1984), 6.139 (which Campbell consulted).

57 Wai-Yee Li, "Gardens and Illusions from Late Ming to Early Qing", *Harvard Journal of Asiatic Studies* 72, no. 2 (2012): 328. The emphasis is mine. The original text can be found in Qi Biaojia, *Qi Biaojia ji* (Beijing: Zhonghua shuju, 1960), 7.167.

58 Wen Fong, *Beyond Respresentation: Chinese Painting and Calligraphy 8th–14th Century* (New York: The Metropolitan Museum of Art, 1992), 85. For the passage in question, see Guo Xi and Guo Si, *Linquan gaozhi,* in *Meishu congshu* 17, ed. Huang Binhong and Deng Shi (Hangzhou: Zhejiang renmin meishu chubanshe, 2013), 17.

59 Mitrović, "Leon Battista Alberti," 425, 429.

60 Pierre Ryckmans, "The Chinese Attitude Towards the Past," *Papers on Far Eastern History* 39 (March 1989): 8.

61 Sun Yirang (1848–1908), *Zhouli zhengyi* (Beijing: Zhonghua shuju, 1987), 83.3423–3430. My translation is based on the one by Paul Wheatley, *The Pivot of the Four Quarters: A Preliminary Enquiry into the Origins and Character of the Ancient Chinese City* (Edinburgh: Edinburgh University Press, 1971), 411. For a French translation, see Edouard Biot, *Le Tcheou-li, ou Rites des Tcheou*, tome II (Paris: Imprimerie nationale, 1851), 556. The *Records on Investigating Crafts* is a document of normative and prescriptive rather than historical nature, formulated during the period of Han Wudi (r. 141–87 BCE) on the basis of a text of the late Warring States period.

62 For a discussion of the employment of mapping and touring modes of description in garden essays, see Craig Clunas, *Fruitful Sites: Garden Culture in Ming Dynasty China* (London: Reaktion Books, 1996), 140–144; Yinong Xu, "Boundaries, Centres, and Peripheries in Chinese Gardens—A Case of Suzhou in the Eleventh Century," *Studies in the History of Gardens and Designed Landscape* 24, no. 1 (March 2004): 31–32.

63 Zhu Changwen, "Lepu ji," in Fan Chengda (1126–1193), *Wujun zhi* (Nanjing:

Jiangsu guji chubanshe, 1986), 14.194–196. For an analysis of the textually presented layout of the Pleasure Patch, see Xu, "Boundaries," 30–33.

64 It is also worth noting that the textual relations between these four domains are quite different in the "Record of the Hall Surrounded by Jade of Master Sitting-in-Reclusion," written by Yuan Huang (1533–1606), from their pictorial adjacency: descriptions of the front court of the "Hall Surrounded by Jade" and the "Remaining Splendour of the Orchid Pavilion" appear very early in the text, quite distantly separated from the description of the remaining two around the two-thirds of the text, and the sequential order of these remaining two is reversed as compared to the succession of their depictions in the scroll. See Yuan Huang, "Zuoyin xiansheng Huancuitang ji," in *Zuoyin xiansheng dingpu*, ed. Wang Tingna (late 16th-early 17th centuries) (Huancuitang edition, ca. 1609), *paobu*, 35a–40a. For a translation of this text, see Alison Hardie, "Record of the Hall Surrounded by Jade of Master Sitting-in-Reclusion," in *The Dumbarton Oaks Anthology of Chinese Garden Literature*, ed. Alison Hardie and Duncan M. Campbell (Washington D.C.: Dumbarton Oaks, 2020), 381–392.

65 Massimo Scolari, *Oblique Drawing: A History of Anti-Perspective* (Cambridge, MA: The MIT Press, 2012), 348.

66 John Hay, "Chinese Space in Chinese Painting," in *Recovering the Orient: Artists, Scholars, Appropriations*, ed. Andrew Gerstle and Anthony Milner (Amsterdam: Harwood Academic Publishers, 1994), 151.

注　释

序

1　*Ideas of Chinese Gardens: Western Accounts, 1300–1860*, Bianca Maria Rinaldi 编辑（Philadelphia, PA: University of Pennsylvania Press, 2016 年），是一部有帮助的选集，较好地汇总了不同时期在欧美出现的关于中国园林的文字。

2　关于这些变化在分裂割据的那几个世纪的清晰描述，见 Donald Holzman, *Landscape Appreciation in Ancient and Early Medieval China: The Birth of Landscape Poetry*, 收录于同作者的 Chinese Literature in Transition from Antiquity to the Middle Ages（Aldershot, England: Ashgate Publishing Co., 1998 年）。

3　比如可以参见周维权，"魏晋南北朝园林概述"，载于《建筑史论文集 第六辑》，清华大学建筑系编辑（北京：清华大学出版社，1984 年），第 80–95 页；作者同上，《中国古典园林史》第二版（北京：清华大学出版社，1999 年），第 81–120 页；王毅，《园林与中国文化》（上海：上海人民出版社，1990 年），第 76–108 页。

4　其他一些学者也把"卧游"译作 "dreaming journey"（梦游）。

5　沈约（441—513 年）等，《宋书》（北京：中华书局，1974 年），第 93 卷，第 2279 页。张彦远，《历代名画记》，俞剑华注释（上海：上海人民美术出版社，1964 年），第 129 页。

6　郭熙（约 1000—约 1087 年）与郭思（1082 年进士），《林泉高致》，载于《美术丛书》，黄宾虹、邓实主编（杭州：浙江人民美术出版社，2013 年），第 2 集，第 7 辑，第 6 页。我的翻译遵循 Wen Fong（方闻），"Monumental Landscape Painting," 载于 *Possessing the Past: Treasures from the National Palace Museum, Taipei*, Wen C. Fong、James C.Y. Watt 编辑（New York: The Metropolitan Museum of Art, 1996 年），第 131 页。

7　Osvald Sirén, *Gardens of China*（New York: Ronald Press Company, 1949 年），第 4 页。喜龙仁马上限定这一观点："当然，我们不能过于强调山水画与园林作品的比较，因为这是两种艺术形式的问题，其中不同的表达媒介差异很大，但这种比较仍然很有价值，并说明了中国人对园林作品的态度。"

8　在此我借用牟复礼的美丽措辞，"我们关注的是一个充满生机的中国，而不是一个垂死的中国。"见 F.W. Mote, "A Millennium of Chinese Urban History: Form, Time, and Space Concepts in Soochow," *Rice University Studies* 59, no. 4（1973 年秋季）：第 39 页。

9　Craig Clunas, *Fruitful Sites: Garden Culture in Ming Dynasty China*（London: Reaktion Books, 1996 年）。

10　画意的英文翻译通常是 "ideas in painting" 或者 "picture–idea"，但是我倾向于遵循约翰·海伊（John Hay）的观点："意"字"经常是更多是'意图'的问题，涉及情感的方向性，而且非常清楚地是欲望的向量。”见 John Hay, “Boundaries and Surfaces of Self and Desire in Yuan Painting,” 载于 *Boundaries in China*, 编辑同上（London: Reaktion Books, 1994 年），第 162 页。同时参见 John Hay, “Some Questions Concerning Classicism in Relation to Chinese Art,” *Art Journal 47*, issue 1（1988 年）：第 32 页。

11　Kristina Kleutghen, “From Science to Art: The Evolution of Linear Perspective in Eighteenth-Century Chinese Art,” 载于 *Qing Encounters: Artistic Exchanges between China and the West,* Petra Ten-Doesschat Chu、Ning Ding 编辑（Los Angeles, CA: The Getty Research Institute, 2015 年），第 173 页。

生活在六朝山水间

1　乐史（930—1007 年），《太平寰宇记》卷九十引《金陵图经》，载于《日本宫内厅书陵部藏宋元版汉籍选刊》，第 58 册（上海：上海古籍出版社，2013 年），第 375 页。

2　陈亮（1143—1194 年），《戊申再上孝宗皇帝书》，载于陈沂（1469—1538 年），《金陵古今图考》（南京：南京出版社，2006 年），第 74 页。

3　范晔（398—445 年），《后汉书》卷八十三《逸民列传》（北京：中华书局，1965 年），第 2756 页。

4　陈寿（233—297 年），《三国志》卷四十八《吴志・孙休传》（北京：中华书局，1959 年），第 1158 页。

5　刘义庆（403—444 年），《世说新语》（上海：上海古籍出版社，1982 年），“言语”第二，第 31 条，第 22 页。

6　同上，第 73 条，第 32 页。

7　同上，“文学”第四，关于清谈有 60 余条，据此总结。

8　同上，第 55 条，第 19 页。

9　同上，第 61 条，第 29 页：“简文入华林园，顾谓左右曰，会心处不必在远，翳然林水，便自有濠、濮间想也。觉鸟兽禽鱼，自来亲人。”

10　许嵩（活跃于公元 8 世纪中叶），《建康实录》卷十二（北京：中华书局，1986 年），第 444 页。

11　何尚之（382—460 年），《华林清暑殿赋》，载于欧阳询（557—641 年）编《艺文类聚》卷六十二（上海：上海古籍出版社，1965 年）：第 1125 页。

12　王融（467—493 年），《三月三日曲水诗序》，载于《艺文类聚》卷四，第 72 页。

13　房玄龄（579—648 年）等，《晋书・谢安传》（北京：中华书局，1974 年），第 2075-2076 页。

14　李延寿（活跃于公元 7 世纪中叶），《南史》（北京：中华书局，1975 年）：列传，卷七十五和卷七十六载“隐逸”人物 31 人，卷七十一载“儒林”人物 19 人。

15　房玄龄等，《晋书・隐逸传》，第 2440 页。

16　萧子显（487—537 年），《南齐书》卷二十二《豫章文献王传》（北京：中华书局，1972 年），第 414 页；姚思廉（557—637 年），《陈书》卷二十五《孙瑒传》（北京：中华书局，1972 年），第 321 页；沈约（441—513 年），《宋书》卷六十七《谢灵运传》（北京：中华书局，1974 年），第 1754 页。

17　左思（约 250—305 年），《招隐》，载于萧统（501—531 年）等辑《昭明文选》卷二十二（郑州：中州古籍出版社，1990 年），第 297 页。

18　沈约，《宋书》卷十五，第 386 页。

19　杜宝（活跃于公元 7 世纪）撰，辛德勇辑校，《大业杂记辑校》（西安：三秦出版社，2006 年），第 49 页。

可行、可望、可游、可居之山水

1　Törmä, Minna, *Enchanted by Lohans: Osvald Sirén's Journey into Chinese Art* (Hong Kong: Hong Kong University Press, 2013 年)

2　Osvald Sirén, *Kinas trädgårdar och vad de betytt för 1700-talets Europa*, 2 Vols.（Stockholm: Svenska Litteratur, 1948 年）。这两卷书的英文版为 *Gardens of China*（New York: Ronald Press, 1949 年）和 *China and the Gardens of Europe of the Eighteenth Century*（New York: Ronald Press, 1950 年）。

3　Ping Foong, *The Efficacious Landscape: On the Authorities of Painting at the Northern Song Court*（Cambridge, MA and London, England: Harvard University Asia Center, 2015 年）。

4　郭熙，《林泉高致》，载于俞剑华编《中国画论类编》，卷一（北京：人民美术出版社，1957 年），第 632-633 页；Susan Bush 和 Hsio-yen Shih 编著 *Early Chinese Texts on Painting*（Cambridge, MA and London: Harvard University Press, 1985 年），其中第 4 章 *The Landscape Texts* 翻译了郭熙的一部分文字，详见第 141-190 页。

5　卜寿珊和时学颜的书讨论了宗炳。详见前注 Bush and Shih, *Early Chinese Texts*，第 36-38 页。

6　郭熙，《林泉高致》，第 632-633 页；Bush and Shih, *Early Chinese Texts*，第 151-152 页。

7　载于 Ju Hsi Chou, *Silent Poetry: Chinese Painting from the Collection of the Cleveland Museum of Art*（Cleveland: Cleveland Museum of Art, 2015 年），第 34-43 页。

8　我在 Traveller 这个词里使用大写的 T 来区别这位观者和他遇见的路人。（译注：中文翻译中使用带引号的"旅行者"来表达这个意思。）

9　传燕文贵《江山楼观图》，手卷，纸本设色，32 厘米 ×161 厘米，大阪市立美术馆。我的分析详见于 Minna Törmä, *Landscape Experience as Visual Narrative: Northern Song Landscape Handsrolls in the Li Cheng—Yan Wengui Tradition*（Helsinki: Academia Scientiarum Fennica, 2002 年）。

10　这两段由葛瑞汉翻译；详见：*Two Chinese Philosophers*, 新编（La Salle, IL: Open Court, 1992 年），第 15 页。

11　我在这里引用的是瑞典文版。Sirén, *Kinas trädgårdar*, Vol. I, 10。这种类比因其

具有园林设计和绘画的普遍关联而在中国园林写作中变得很常见。有关这种方法的评论，详见 Jing Xiao and Charlie Q. L. Xue. "Architecture in Ji Cheng's *The Craft of Gardens*: a visual study of the role of representation in counteracting the influence of the pictorial idea in Chinese scholar gardens of the Ming period," *Studies in the History of Gardens & Designed Landscape* 35, no. 3（2015 年），第 218–234 页。

12 Sirén, *Kinas trädgårdar*, Vol. I, 第 103 页。

13 本文使用的网师园照片摄于 2009 年 6 月。但是拍摄的时候并没有根据序列描述来选择拍摄点。同时期在留园拍摄的照片还增加了大量的细部。

14 这些景点的译名来自：Alfreda Murck and Wen Fong, *A Chinese Garden Court: The Astor Court at the Metropolitan Museum of Art*（New York: Metropolitan Museum of Art, 1985 年），第 11 和 14 页。

15 手卷，绢本设色，35.6 厘米 ×104.4 厘米。北京故宫博物院。详见 http://en.dpm.org.cn/collections/collections/2013-01-24/1207.html（2018 年 10 月 21 日访问）。

16 巫鸿挖掘了这种现象的方方面面。详见：Wu Hung, *The Double Screen*（London: Reaktion Books, 1996 年）。

17 手卷，绢本设色，29.5 厘米 ×71.1 厘米。肯萨斯城纳尔逊-阿特金斯艺术博物馆。详见 https://art.nelson-atkins.org/objects/10336/whiling-away-the-summer-hsiaohsia-tu?ctx=00292d92-75aa-4034-9884-8d990f29808b&idx=0（2018 年 10 月 21 日访问）。《重屏会棋图》有很多版本，本文引用的是传为周文矩（活跃于十世纪中叶）作、实乃明代十四世纪仿作的一幅；手卷，水墨设色绢本，31.3 厘米 ×50 厘米；佛利尔与赛克勒美术馆（Charles Lang Freer 赠品）。https://www.freersackler.si.edu/object/F1911.195/（2018 年 10 月 21 日访问）。

18 大都会美术馆中的园林庭院（明轩）仿建网师园中的庭院之一（图 6–图 7）。这本书详述了网师园的历史和结构以及明轩的建造。

19 Chen Congzhou, *On Chinese Gardens*（New York: Better Link Press, 2008 年），第 15 页。

20 Murck and Fong, *Chinese Garden Court*，第 30 页。文字强调由我所加。

21 肖靖和薛求理注意到"图画美学似乎在某种程度上抑制了眼睛的生理运作。"见 Jing and Xue, "Architecture in Ji Cheng's," 第 218 页。

22 芝加哥艺术博物馆收藏这一系列中的一幅：https://www.artic.edu/artworks/79314/dream-journey-to-rivers-and-mountains (2018 年 11 月 21 日访问)。

23 Li Wai-yee, "Gardens and Illusions from Late Ming to Early Qing," *Harvard Journal of Asiatic Studies* 72, no. 2（2012 年 12 月）：第 295–336 页。

24 同上，第 298–299 页。

25 同上，第 298–302 页。"乌有园"来自刘士龙（十七世纪），"意园"来自戴名世（1653—1713 年）。

26 Ink and colours on paper, 116.6 × 23.8 cm; Art Institute of Chicago；见：https://www.artic.edu/artworks/40558/tea-drinking-under-the-wutong-tree?q=tang+yin（2018 年 11 月 27 日访问）。也见于 https://scrolls.uchicago.edu/view-scroll/184（2018 年 11 月 27 日访问），在这个网站上可以实现数字化展卷。

景冶：中国园林中的造景技艺

1　虽然从词源学上讲“景观（spectacle）”一词仅和视觉有关，但这里的使用更为广义，指那些能吸引观者各种感官之物。

2　关于住宅和园林间的相互作用，参见：Antoine Gournay. *La Maison chinoise*（Paris: Klincksieck, 2016 年）。

3　可参见：Edwin T. Morris, *The Gardens of China: History, Art and Meanings*（New York: Charles Scribner's Sons, 1983 年）。杨鸿勋，《江南园林论：中国古典造园艺术研究》（上海：上海人民出版社，1994 年）。

4　关于这个术语具有的多种可能含义，参见 Duncan Campbell 在 “China Heritage Glossary,” *China Heritage Quarterly*, no. 29（March 2012）中的“景”词条，在线网址为：http://www.chinaheritagequarterly.org/glossary.php?searchterm=029_jing.inc&issue=029

5　参见 Pauline Yu, *The Poetry of Wang Wei: New Translations and Commentary*（Bloomington: Indiana University Press, 1980 年）。Patrick Carré, *Les saisons bleues:* L'*œuvre de Wang Wei poète et peintre*（Paris: Phébus, 1989 年）。

6　William Chambers, *A dissertation on Oriental Gardening*（London: Griffin, 1772 年），第 22 页。

7　Antoine Gournay. “Chine: jardins du Lingnan à la fin de la dynastie des Qing (1644–1911),” *Polia, Revue de l'art des jardins*, n° 1（printemps 2004 年）：第 63–78 页。

8　关于这些方面的全面介绍，参见 Antoine Gournay. “Pour une analyse médiationniste du jardin chinois,” *Tétralogiques*, n° 23（2018 年），*Le modèle médiationniste de la technique*. http://tetralogiques.fr/spip.php?article93。

9　北京市园林绿化局编，《北京古树名木》（北京：燕山出版社，2009 年），第 223 页。

10　Terese Tse Bartholomew, *Hidden meanings in Chinese Art*（San Francisco: Asian Art Museum of San Francisco, 2006 年），第 7 章，第 41，47–48 页。

11　此处用“观法（spectation）”来和“景观（spectacle）”进行区分。这个概念借用自：Ph. Bruneau, “Qu'est-ce qu'une église?” *Ramage-Revue d'archéologie moderne et d'archéologie générale*, n° 9（1991），第 49–70 页。其在中国园林研究中的应用，参见：Antoine Gournay, “Le jardin chinois,” *Ramage*, n° 12（1994—1995 年）：第 119–135 页。

12　Jean-Denis Attiret. “Lettre à M. d' Assaut, 1er novembre 1743,” in *Lettres édifiantes et curieuses écrites des missions étrangères*（Paris, 1747 年），第 28 卷，第 1–49 页。英文翻译见 Sir Harry Beaumont, *A Particular Account of the Emperor of China's Gardens Near Pekin*（London: Dodsley, 1752 年），第 8 页。https://archive.org/details/aparticuracco00attigoog/page/n23。

13　Chambers, *Dissertation on Oriental Gardening*，第 64 页。

14　Ji Cheng, *The Craft of Gardens*, Alison Hardie 翻译（New Haven and London: Yale University Press, 1988 年），第 43 页。

15　关于盆景艺术及其与道家思想的联系，参见：Rolf A. Stein, “Jardins en miniature d' Extrême-Orient,” *Bulletin de l' École Française d' Extrême-Orient*, n° 42（1943

年）：第 1-104 页；修改更新于 *Le Monde en petit: Jardins en miniature et habitations dans la pensée religieuse d'Extrême-Orient*（Paris: Flammarion, 1987 年）。英文翻译版：*The World in Miniature: Container Gardens and Dwellings in Far Eastern Religious Thought*（Stanford: Stanford University Press, 1990 年）。

16　关于这一术语在《园冶》中的使用，参见：Wybe Kuitert, "Borrowing scenery and the landscape that lends—the final chapter of Yuanye," *Journal of Landscape Architecture* 第 10 卷，第 2 期（2015），第 32-43 页。网址：https://www.researchgate.net/publication/279245717_Borrowing_scenery_and_the_landscape_that_lends_-_The_final_chapter_of_Yuanye

17　胡东初，《中国文化与园林艺术》（北京：新世界出版社，1991 年），第 93-95 页。关于墙体开洞所扮演的角色方面的讨论详见：Antoine Gournay, "Le système des ouvertures dans l'aménagement spatial du jardin chinois," *Extrême-Orient, Extrême-Occident, n° 22, L'art des jardins dans les pays sinisés, Chine, Japon, Corée, Vietnam*, I Léon Vandermeersch 编辑（2000 年）：第 51-71 页。

18　英译来自 Alison Hardie，见 Ji Cheng, *Craft of Gardens*，第 76 页。

19　王毅，《园林与中国文化》（上海：上海人民出版社，1990）。罗豫所译英文版为：Wang Yi, *A Cultural History of Classical Chinese Gardens*（New York: SCPG Publishing Corporation, 2015 年），其中图 1-图 8 为上海秋霞圃同题场景。

"知夫画脉"与"如入岩谷"：清初寄畅园的山水改筑与十七世纪江南的"张氏之山"

1　原文刊载于：顾凯，"'知夫画脉'与'如入岩谷'：清初寄畅园的山水改筑与 17 世纪江南的'张氏之山'"，《中国园林》总 35 期（2019 年第 7 期）：第 124-129 页。本文中有少许改动。

2　如朱有玠说："我在诸多明清遗存的江南名园中，最赞赏无锡的寄畅园"，见秦志豪主编，《锡山秦氏寄畅园文献资料长编》（上海：上海辞书出版社，2009），"题词"；潘谷西主编的《中国建筑史》教材，已历七版，寄畅园一直为"明清江南私家园林"案例之首，见潘谷西编著，《中国建筑史》第七版（北京：中国建筑工业出版社，2015）；曹汛称他评选中国的"五个金牌园林"，排第一的就是寄畅园，见曹汛，"江南园林甲天下，寄畅园林甲江南"，《风景园林》总 25 期（2018 年 11 期）：第 14 页；董豫赣则"许无锡寄畅园为生平所见第一"，见董豫赣，"双园八法：寄畅园与谐趣园比对"，《建筑师》总 172 期（2014 年 12 月）：第 108 页；等等。

3　黄晓，"锡山秦氏寄畅园考"（北京大学硕士论文，2010），第 96 页。

4　徐永言修，秦松龄、严绳孙纂，《无锡县志》康熙二十九年刻本，卷七，第 9b 页。

5　黄晓，刘珊珊，"明代后期秦燿寄畅园历史沿革考"，载于贾珺主编，《建筑史》第 28 辑（北京：清华大学出版社，2012 年），第 112-135 页。

6　黄晓，"锡山秦氏寄畅园考"。

7　黄晓，刘珊珊，"寄畅园：一座园林背后的文明轮回"《中华遗产》总 104 期（2014 年 6 月）：第 46-65 页。

8　秦志豪，"清初寄畅园的归并和改筑"，载于《寄畅园建园 490 周年研讨会论文集》（无锡，2017 年），第 1-9 页。

9 毛茸茸，“与君犹对秦楼月：惠山秦氏寄畅园研究”（中国美术学院博士论文，2016年），第97页。

10 徐永言修，秦松龄、严绳孙纂，《无锡县志》，卷七，第9b页。

11 曹汛，“张南垣的造园叠山作品”，载于王贵祥主编《中国建筑史论汇刊》第2辑（北京：清华大学出版社，2009），第327-378页。

12 同上，第374页。

13 同上，第327页。

14 顾凯，“重新认识江南园林：早期差异与晚明转折”，《建筑学报》（2009年S1期）：第106-110页。

15 曹汛，“造园大师张南垣（一）：纪念张南垣诞生四百周年”，《中国园林》（1988年第1期）：第21-26页。

16 吴定璋编，《七十二峰足征集·文集》乾隆十年刻本，卷十，第14a页。

17 曹汛，“造园大师张南垣（二）：纪念张南垣诞生四百周年”，《中国园林》（1988年第3期）：第5-6页。

18 顾凯，“画意原则的确立与晚明造园的转折”，《建筑学报》（2010年S1期）：第127-129页。

19 吴伟业，《吴梅村全集》（上海：上海古籍出版社，1990年），第1059-1061页。

20 顾凯，“拟入画中行：晚明江南造园对山水游观体验的空间经营与画意追求”，《新建筑》（2016年第6期）：第44-47页。

21 戴名世，《南山集》，载于《续修四库全书·集部·一四一九》（上海：上海古籍出版社，2002年），第143页。

22 曹汛，“造园大师张南垣（二）”，第4页。

23 顾凯，“拟入画中行”，第44-47页。

24 黄宗羲，《黄宗羲全集》第10册《南雷诗文集上》（杭州：浙江古籍出版社，2012年），第571页。

25 邵长蘅，《邵子湘全集》，载于《四库全书存目丛书·集部二四七》（济南：齐鲁书社，1997年），第715页。

26 秦志豪主编，《锡山秦氏寄畅园文献资料长编》（上海：上海辞书出版社，2009年），第82页。

27 顾凯，《明代江南园林研究》（南京：东南大学出版社，2010年），第183-184页。

28 关于清初寄畅园改筑的具体时间，目前有两种推断：黄晓推定“在康熙六至七年间”（“锡山秦氏寄畅园考”，第96页），秦志豪则判断“从康熙五年春到康熙六年秋”（“清初寄畅园的归并和改筑”，第7页），无论何者都认为总体用时不超过两年。

29 秦志豪主编，《锡山秦氏寄畅园文献资料长编》，第37页。

30 梁洁等通过对晚明寄畅园考古材料的分析得出结论：“晚明的寄畅园确乎可

称得上是一座豪华的园林。”参见：梁洁、郑炘，“晚明寄畅园水池‘锦汇漪’及其周边复原研究”，《中国园林》总34期（2018年第12期）：第138页。

31 谈修，《惠山古今考》明万历间刻本，卷一，第17a页。

32 邵长蘅，《邵子湘全集》，第715页。

33 顾凯，“中国传统园林中‘亭踞山巅’的再认识：作用、文化与观念变迁”，《中国园林》总32期（2016年第7期）：第78-83页。

34 秦志豪主编，《锡山秦氏寄畅园文献资料长编》，第126页。

35 同上，第131页。

36 同上，第129页。

37 邵长蘅，《邵子湘全集》，第715页。

38 鹤步滩之树在清中期的各图像中即可明显看到，并非只是今日景象。

39 在康熙十五年吴绮《寄畅园杂咏》九首中其八为“宛转桥”（吴绮，《林蕙堂全集》文渊阁四库全书本，卷十五臯诗集，43b），在《清高宗南巡名胜图》和钱维城《弘历再题寄畅园诗意卷》中现在“七星桥”的位置都是三折曲桥，可见早期并非如现在的直桥。另外，清中期图像中都尚未出现今日水池东北角的跨水廊桥。

40 秦志豪主编，《锡山秦氏寄畅园文献资料长编》，第88页。

41 如董豫赣所言：“寄畅园池形的狭长，一笔而两得山高水远—以其狭长之狭，横对逼山则山高；以其狭长之长，纵观水长则水远。”参见：董豫赣，“双园八法—寄畅园与谐趣园比对”，《建筑师》（2014年第6期）：第108-117页。

42 秦志豪主编，《锡山秦氏寄畅园文献资料长编》，第10页。

43 同上，第24页。

44 同上，第35页。

45 同上，第33页。

46 同上，第34页。

47 同上，第42页。

48 清初改筑后的文献中仍多有瀑景提及，如余怀《寄畅园宴集放歌》之“忽闻瀑布声潺潺”、吴绮《寄畅园杂咏·石径》之“飞瀑四时秋”、清高宗《寄畅园》之“不尽烟霞飞瀑潺”等；清中期的图像中也可见入池之瀑的形象。

49 许缵曾，《宝纶堂稿》，载于《续修四库全书·集部·一四〇九》（上海：上海古籍出版社，2002年），第651页。

50 秦志豪主编，《锡山秦氏寄畅园文献资料长编》，第81页。

51 同上，第92页。

52 秦志豪，“清初寄畅园的归并和改筑”，第6页。

53 顾凯，“明末清初太仓乐郊园示意平面复原探析”，《风景园林》总139期（2017年第2期）：第25-33页。

54 “小阮”即侄儿。魏末晋初“竹林七贤”中的阮籍与阮咸为叔侄，世称阮籍为大阮，阮咸为小阮，后人便以“大小阮”作为叔侄关系的代称。

55 徐永言修，秦松龄、严绳孙纂，《无锡县志》，卷七，第 9b 页。

56 李念慈撰，《谷口山房诗集》，载于《四库全书存目丛书·集部二三二》（济南：齐鲁书社，1997 年），第 794 页。

57 秦志豪主编，《锡山秦氏寄畅园文献资料长编》，第 85 页。

58 同上，第 120 页。

59 当代此涧名为“八音涧”，人们往往认为此名称自清初即有，然而遍查寄畅园史料，“八音涧”的称谓要到 20 世纪的民国整修时才出现。

60 吴绮，《林蕙堂全集》文渊阁四库全书本，卷十五，第 43b 页。

61 秦志豪主编，《锡山秦氏寄畅园文献资料长编》，第 123 页。

62 秦志豪，“清初寄畅园的归并和改筑”，第 6 页。

63 秦志豪主编，《锡山秦氏寄畅园文献资料长编》，第 84 页。

64 陈从周著，《说园》（上海：同济大学出版社，1984 年），第 40 页。

真实的园林与想象的园林：园林、山水画、别号图

1 关于这时期中国艺术中的“礼物”，参见：Craig Clunas, *Elegant Debts: The Social Art of Wen Zhengming*（London: Reaktion Books, 2004 年）。中译本：柯律格，《雅债：文徵明的社交性艺术》（北京：生活·读书·新知三联书店，2012 年）。

2 张丑，《清河书画舫》，作于 1616，《钦定四库全书》版，卷十一，第 42b 页，见 https://ctext.org/library.pl?if=gb&res=6240 (2018 年 10 月 29 日访问）。文中重点为笔者所加。

3 关于明代的园林文化，参见：Craig Clunas, *Fruitful Sites: Garden Culture in Ming Dynasty China* (London: Reaktion Books, 1996 年)。

4 柯律格，《雅债》，第 99 页；引自（明）文徵明著，周道振编，《文徵明集》（上海：上海古籍出版社，1987 年），第 502 页。

5 刘九庵，“吴门画家之别号图及鉴别举例”《故宫博物院院刊》，总第 49 期（1990 年第三期）：第 54-61 页。

6 许珂，“君子于室：文徵明别号图中的图解与隐喻”《美术学报》（2016 年第 5 期）：第 33-40 页；许珂，“吴门画派别号图的文化意义与社会功用：以文徵明别号图为例”《南京艺术学院学报（美术与设计）》（2016 年第 6 期）：第 55-61 页。

7 http://www.dpm.org.cn/collection/paint/228141.html?hl=%E5%8F%8B%E6%9D%BE%E5%9C%96（2018 年 10 月 12 日访问）。

8 张丑，《清河书画舫》，卷十二上，第 7b 页。

9 全卷可参看：https://m-minghuaji.dpm.org.cn/paint/appreciate?id=08f185a3d4e3467cb073832559f59447（2020 年 10 月 12 日访问）。

10　《文徵明集》，第 582 页，转引自：Clunas, *Elegant Debts*，第 44 页。

11　Richard Edwards, *The Art of Wen Cheng-ming (1470–1559)*（Ann Arbor MI: University of Michigan Museum of Art, 1976 年），第 138–139 页。

12　全卷可参看：https://m-minghuaji.dpm.org.cn/paint/appreciate?id=41ebf4498dfa4d58b89705e4d8dcf8a4（2020 年 10 月 12 日访问）。

13　刘九庵，“吴门画家之别号图及鉴别举例”，第 58 页。

14　计成原著，陈植注释，《园冶注释》第 2 版（北京：中国建筑工业出版社，1988 年），第 47 页。

15　吴伟业著，李学颖集评标校，《吴梅村全集》（上海：上海古籍出版社，1990 年），第 1060 页。

16　我用“Man of the Way”而非“Daoist”来翻译“道人”，因为此时吴历正在皈依天主教过程中，这是与中国传统哲学不同的“道”。

17　http://www.dpm.org.cn/collection/paint/228483.html?hl=%E6%B1%82%E5%BFF%97%E5%9C%92（2018 年 10 月 12 日访问）。

18　https://insitu.arthistory.wisc.edu/Details/10462/269909（2020 年 10 月 12 日访问）。关于文徵明与张凤翼、张献翼兄弟的关系，参见：Clunas, *Elegant Debts*，第 130 页。

19　王世贞，《弇州山人四部稿》，成书于 1577 年（台北：伟文图书出版社，1976 年），第 3581 页。

20　http://painting.npm.gov.tw/Painting_Page.aspx?dep=P&PaintingId=4016（2018 年 10 月 12 日访问）。

21　王世贞，“弇山园记”，载于《弇州山人续稿》。参见：沈云龙编，《明人文集丛刊》（台北：文海出版社，1970），第 2957–2997 页。也可参见：陈植、张公弛编，《中国历代名园记选注》（合肥：安徽科学技术出版社，1983），第 130–156 页。

22　Kate Kerby and Mo Zung Chung, *An Old Chinese Garden: A Three-fold Masterpiece of Poetry, Calligraphy and Painting*（Shanghai: Chung Hwa Book Co., 1922）。网络的翻拍图片可参见：http://blog.sina.com.cn/s/blog_647cbe2b0100h6kq.html（2018 年 10 月 18 日访问）；槐雨亭在接近底部位置。

23　文徵明著，周道振辑校，《文徵明集》（上海：上海古籍出版社，1987 年），第 1276 页。

24　文徵明著，《文徵明集》，第 1210 页。

25　https://www.shanghaimuseum.net/museum/frontend/collection/zoom.action?cpInfoId=949&picId=1863（2018 年 10 月 12 日访问）。

26　文徵明著，《文徵明集》，第 1303 页。

27　http://www.nipic.com/show/2/27/4803652k08eec23e.html（2018 年 10 月 12 日访问）。

28　英文译文参见：Edwards，*The Art of Wen Cheng-ming*，第 150–151 页；其中也有彩色复制品。也可参看：https://www.metmuseum.org/art/collection/search/39536（2020 年 10 月 12 日访问）。

29　故宫博物院网站暂无此画。彩色印刷图版（含钱氏跋文）参见：柳如是撰，周书田、范景中辑校，《柳如是集》（杭州：中国美术学院出版社，2002 年）。

30 James Cahill, *The Distant Mountains: Chinese Painting of the Late Ming Dynasty, 1570–1644*（New York & Tokyo: Weatherhill, 1982 年），第 132 页。故宫博物院网站暂无此画。此画可参见：http://auction.artron.net/paimai-art00270113/（2018 年 10 月 23 日访问）。

31 Cahill, *Distant Mountains*，第 39–40 页。

32 James Cahill, *The Compelling Image: Nature and Style in Seventeenth-Century Chinese Painting*（Cambridge, MA and London, England: Harvard University Press, 1982 年），第 6–7 页。

33 http://www.dpm.org.cn/collection/paint/229436.html（2018 年 10 月 18 日访问）。

34 http://www.dpm.org.cn/collection/paint/228964.html?hl= 錢穀（2018 年 10 月 18 日访问）。

35 曹汛，"清代造园叠山艺术家张然和北京的'山子张'"，载于《建筑历史与理论》第二辑，中国建筑学会、建筑历史学术委员会编（南京：江苏人民出版社，1981 年），第 116 页。

36 完整画册的彩图色版参见：June Li and James Cahill, *Paintings of Zhi Garden by Zhang Hong: Revisiting a Seventeenth-Century Chinese Garden*（Los Angeles: Los Angeles County Museum of Art, 1996 年）。全景视角的黑白图版参见：Cahill, *Compelling Image*, 第 20 页。也可参看：http://www.zhigarden.com/PicDetail.aspx?ID=181（2020 年 10 月 12 日访问）。

37 Cahill, *Compelling Image*，尤其是第三章；关于项圣谟的部分见第 98–105 页。

38 Cahill, *Distant Mountains*，第 86 页。

39 赵晓华，"史有别号入画来：唐寅《悟阳子养性图》考辨"，《文物》（2000 年第 8 期）：第 78–83 页。

40 韩金磊，"唐寅'别号图'创作中'景'与'人'的组构关系"，《翠苑（民族美术）》（2016 年第 7 期）：第 76–78 页。

41 叶玉，"别号图与山水画的象征意义：从文徵明的《真赏斋图》谈起"，《新美术》（2016 年第 12 期）：第 16–23 页。

胸中丘壑：一方没有"空间"的天地

1 刘敦桢，"苏州的园林"，《南京工学院院刊》第四期单行本（1957 年）：第 8、11、12、15 页。

2 刘敦桢，《苏州古典园林》（北京：中国建筑工业出版社，1979 年），第 9–10 页。该卷由潘谷西编纂。

3 童寯在其 "Architecture Chronicle" 中不避免谈及空间，见《天下月刊》(*T'ien Hsia Monthly*) V, no. 3（1937 年 10 月）：第 308–312 页。我参阅的是《童寯文集》第一卷（北京：中国建筑工业出版社，2000 年），第 81–88 页。

4 童寯，"Chinese Gardens: Especially in Kiangsu and Chekiang"，《天下月刊》

(*T'ien Hsia Monthly*) III, no. 3（1936 年 10 月）：第 220 页："中国园林实非旷地一块，而是分成走廊和庭院，是房屋而非植物在那里起支配作用。"在这一语境下，方拥、汪坦在童寯的《园论》（天津：百花文艺出版社，2006 年）第 1 页里对"空间"一词的翻译是很准确的。

5　童寯，《江南园林志》，第二版（北京：中国建筑工业出版社，1987 年）。

6　童寯，"Suzhou Gardens: Which Embody All the Characteristics of the Chinese Landscape Art"，载于《童寯文集》第一卷（北京：中国建筑工业出版社，2000 年），第 262-274 页，引用语句在第 269 页；中文版《苏州园林：集中国造园艺术特征于一体》，李大夏、汪坦译，同一出处，第 275-282 页。同样，在 1983 年完成、1997 年出版的"Glimpses of Gardens in Eastern China"（中文版为《东南园墅》。北京：中国建筑工业出版社）中也提到过"空间"。

7　见顾凯，"童寯与刘敦桢的中国园林研究比较"，《建筑师》总 173 期（2015 年 2 月）：第 92-105 页。

8　"作为一个术语"，Adrian Forty 提醒我们，"'空间'直到十九世纪九十年代才出现在建筑学词汇中"。见 *Words and Buildings: A Vocabulary of Modern Architecture*（London: Thames & Hudson, 2000 年），第 256 页。

9　Clifford Geertz, *The Interpretation of Cultures: Selected Essays*（New York: Basic Books, 1973 年），第 5 页。Geertz 在提到"诠释性"问题时，用了"科学"一词。但"科学"在此或为误导，因为文化理论属于人文学科的一部分，在原理和方法上与自然科学迥异。就这一观点而言，"学科"或许更为恰当。但翻译必须尊重原文。

10　F.W. Mote, "The Arts and the 'Theorizing Mode' of the Civilization,"载于 *Artists and Tradition: Uses of the Past in Chinese Culture*, Christian F. Murck 编辑（Princeton, NJ: Princeton University Press, 1976 年），第 5 页。

11　例如，Minna Törmä, "Looking at Chinese Landscape Painting,"载于 *Looking at Other Cultures: Works of Art as Icons of Memory*, Anja Kervanto Nevanlinna 编辑（Helsinki: Taidehistorian Seura, 1999 年），第 119-135 页；鲁安东，"Lost in translation: Modernist Interpretation of the Chinese Garden as Experiential Space and Its Assumptions," *The Journal of Architecture* 16, no. 4（2011 年）: 第 499-527 页。

12　James Elkins, *The Poetics of Perspective* (Ithaca, NY: Cornell University Press, 1994 年), 第 41 页。

13　Erwin Panofsky, *Perspective as Symbolic Form*, Christopher S. Wood 翻译（New York: Zone Book, 1991 年），第 27 页。Branko Mitrović 在这一语境下讨论了空间的同质性问题，见其"Leon Battista Alberti and the Homogeneity of Space," *Journal of the Society of Architectural Historians* 63, no. 4（2004 年 12 月）：第 424-425 页。

14　James Elkins, "Renaissance Perspectives," *Journal of the History of Ideas* 53, no. 2（1992 年 4-6 月）：第 209、217 页。

15　William V. Dunning, *Changing Images of Pictorial Space: A History of Spatial Illusion in Painting*（New York: Syracuse University Press, 1991 年），第 65 页。

16　Marco Musillo, *The Shining Inheritance: Italian Painters at the Qing Court, 1699-1812*（Los Angeles, CA: The Getty Research Institute, 2016 年），第 120-121、126 页，认为另一个关于欧洲与中国绘画经验对立的说法也因"过度简化"而不值得考虑，即欧洲的透视运用被认为是一种引发物质现实的技术，这与中国的寻找绘画中的精神和诗意的传统形成对比。Musillo 举例说明了在早期现代艺术文献中，欧洲绘画中的灵性探寻同样显而易见，不过他用反问句的形式以古希腊艺术概念为基础做出普遍性断言令人颇感奇怪："模仿与创造：这两种人类表达形式之间的相互作用难道不是中国和欧洲绘画传统的共同问题吗？"（第 127 页）我们还应该注意到：一方面，欧洲透视绘画的早期发展，不但与用几何方式再现理想的基督教世界、以便揭示上帝以秩序和逻辑原则为基础创造世界这一想法有关，而且与创作 Edgerton 所说的"为上帝服务的图画"这一想法有关；另一方面，线性透视刚被发明出来时并没有考虑到明确的科学应用，只是为了帮助解决一个典型的中世纪神学问题，即去除观者世界与图像领域之间那种陈旧的偶像障碍，以便上帝和他的圣徒在观者日常生活中再次不可或缺，仿佛他们都是实际存在的人物，真人大小、栩栩如生，此时此地、可视可触。参较 Samuel Edgerton, *The Renaissance Rediscovery of Linear Perspective*（New York: Basic Books, 1975 年），第 16、157 页；Dunning, *Changing Images*，第 58 页；孟德卫（D.E. Mungello），*The Great Encounter of China and the West, 1500–1800*, 第 4 版（Plymouth, England: Rowman & Littlefield, 2013 年），第 40 页；李启乐（Kristina Kleutghen），*Imperial Illusions: Crossing Pictorial Boundaries in the Qing Palaces*（Seattle, WA: The University of Washington Press, 2015 年），第 80 页。

17　关于这些辩论，参较 Panofsky, *Perspective as Symbolic Form*; *Edgerton, Renaissance Rediscovery*；Martin Kemp, *The Science of Art: Optical Themes in Western Art from Brunelleschi to Seurat*（New Haven, CT: Yale University Press, 1990 年），尤其是第 7–162 页；Hubert Damisch, *The Origin of Perspective*, John Goodman 翻译（Cambridge, MA: The MIT Press, 1994 年）；James Elkins, "Renaissance Perspectives" 和 *The Poetics of Perspective*；Mitrović, "Leon Battista Alberti"，第 424–439 页。

18　Norman Bryson, *Word and Image: French Painting of the Ancien Régime*（Cambridge, England: Cambridge University Press, 1981 年），第 89–91 页。这一观点完全被 Dunning 采纳，见 *Changing Images*，第 104 页。

19　Joseph Needham, *Science and Civilisation in China* Volume IV, Part 3（Cambridge, England: Cambridge University Press, 1971 年），第 114–115 页。

20　James O. Caswell, "Lines of Communication: Some 'Secrets of the Trade' in Chinese Painters' Use of 'Perspective'," *RES* 40（2001 年秋季）：第 189、190 页。

21　纵览近些年来关于中、欧艺术互动背景下康乾盛世的艺术与视觉文化方面的学术研究，参见王正华（Cheng-Hua Wang），"Global Perspective on Eighteenth-Century Chinese Art and Visual Culture," *The Art Bulletin* 96, no. 4（2014 年）：第 379–394 页。关于十八世纪中国艺术中线性透视的条件和发展的简要讨论，参见李启乐（Kristina Kleutghen），"From Science to Art: The Evolution of Linear Perspective in Eighteenth-Century Chinese Art," 载于 *Qing Encounters: Artistic Exchanges between China and the West*，曲培醇（Petra Ten-Doesschate Chu）、丁宁编辑（Los Angeles, CA: The Getty Research Institute, 2015 年），第 173–189 页。

22　关于奥村政信在历史方面的成就，参见 Hugo Munsterberg, *The Japanese Print: A Historical Guide*（New York: Weatherhill, 1982 年），第 37–45 页；Andreas Marks, *Japanese Woodblock Prints: Artists, Publishers and masterworks 1680–1900*（North Clarendon, VT: Tuttle Publishing, 2010 年），第 32–35 页。

23　Timon Screech, "The Meaning of Western Perspective in Edo Popular Culture," *Archives of Asian Art* 47（1994 年）：第 59 页。斯克里奇再次提到同一幅画，他的推测以另一种方式写入 *The Lens within the Heart: The Western Scientific Gaze and Popular Imagery in Later Edo Japan*（Honolulu, HI: University of Hawai'i Press, 2002 年），第 104 页："他最早的作品不是外国深远的透视景象，而是展示日本的各个场所，并且往往是室内场景，这可能是由于艺术家的自我约束，即他还没有足够的自信来表现出不规则的、更加弯曲且参差不齐的户外景象。"

24　Stephen Little, "The Lure of the West: European Elements in the Art of the Floating World," *Art Institute of Chicago Museum Studies* 22, no. 1（1996 年）：第 78–80 页。但是，Harold P. Stern 在简析奥村政信的另一件作品时，以尊重的方式推迟了这类价值判断，

见“A Private Puppet Performance,”载于*Freer Gallery of Art Fiftieth Anniversary Exhibition I. Ukiyo-e Painting*（Washington D.C.: Smithsonian Institution, 1973 年），第 91 页。

25　Screech, *Lens within the Heart*，第 104 页。

26　关于酒吞童子文字的介绍和翻译，参见 Noriko T. Reider, “Shuten Dōji: ‘Drunken Demon’,” *Asian Folklore Studies* 64, no. 2（2005 年）：第 207–231 页。

27　Yoriko Kobayashi-Sato、Mia M. Mochizuki, “Perspective and Its Discontents or St. Lucy’s Eyes,” 载于 *Seeing Across Cultures in the Early Modern World*, Dana Leibsohn、Jeanette Favrot Peterson 编辑（Surrey, England: Ashgate, 2012 年），第 28 页。

28　同上，第 35 页。

29　Angus Lockyer, “Hokusai’s Thought,” 载于 *Hokusai: Beyond the Great Wave*, Timothy Clark 编辑（London, England: Thames & Hudson, 2017 年），第 33 页。

30　Screech, “Meaning of Western Perspective,” 第 66–67 页。

31　Kobayashi-Sato、Mochizuki, “Perspective and Its Discontents,” 第 32、35–36 页。这二位作者引用的是题为《中村座顔见世狂言》的版画。

32　同上，第 36–37 页。

33　同上，第 39 页，图 1.8。

34　Katsushika Hokusai, *Hokusai Manga*（Nagoya: Eirakuya Toshiro 永乐屋东四郎，1815 年），3.15b–16a。大英博物馆免费图片服务，1979,0305,0.428.3。戴维・贝尔（David Bell）认为这一技术是“一件奇特之事”，见 *Ukiyo-e Explained*（Kent, England: Global Oriental，2004 年，第 210 页），而小林頼子和望月みや（“Perspective and Its Discontents,” 第 38–42 页）则暗示透视线聚合轴的透视画法是浮世绘艺术家的一项创新，并援引杉田玄白（Sugita Genpaku）的《解体新书》（1774 年）以强化其论证力度，“实际上是由小田野直武（Odano Naotake）从荷兰版的胡安・德・瓦尔韦德（Juan de Valverde）的 1556 年和 1560 年的《人体解剖学》（*La anatomia del corporis humani*）直接复制的……”。事实却并非如此；具体来说，鱼骨透视技术可能既不是浮世绘艺术家奇特的做法，也不是葛饰北斋的发明。十六世纪中叶欧洲出版的许多关于人体解剖学的书，例如胡安・德・瓦尔韦德的《解剖或零件的活动图像……》（*Anatomie, oft levende beelden vande deelen ...*，1568 年），都有一幅卷首图，其上清楚地显示了透视线聚合轴透视的使用。可以确定的是，浮世绘艺术家欣然采用了这种方法，因为这对他们来说很好用。

35　Kobayashi-Sato、Mochizuki, “Perspective and Its Discontents,” 第 3 页。

36　Anita Chung, *Drawing Boundaries: Architectural Images in Qing China*（Hawai‘i, HI: University of Hawai‘i Press, 2004 年），第 94 页。

37　这种说法可能会产生误导，好像这种体式是造成观看模式的原因一样。尽管本文篇幅有限，无法详细说明这一点，但是我认为，特定的体式和独特的观看模式彼此互为因果。

38　Musillo, *The Shining Inheritance: Italian Painters*，第 89–92 页。

39　Maxwell K. Hearn, “Art Creates History: Wang Hui and *The Kangxi Emperor’s Southern Inspection Tour*,” 载于 *Landscapes Clear and Radiant: The Art of Wang Hui (1632–1717)*, Maxwell K. Hearn 编辑（New York: The Metropolitan Museum of Art, 2008 年），第 180–181 页。

40　Needham, *Science and Civilisation*, 第 112 页。

41 但李约瑟（同上，第 381 页）在评论元朝时期穆斯林对中国天文学的影响时，确实表明“中国人未采用星盘和日晷上的立体投影所需的发达的几何学”，并指出“耶稣教士带来了对行星运动的几何分析的清晰阐述，当然还有其实际应用所必需的欧几里得几何”（同上，第 437 页）。

42 无论是现代到来之前的汉语词典（例如《佩文韵府》和《经济纂诂》），还是专门针对古典汉语的现代汉语词典（例如《辞源》），其中都没有“空间”的条目。此词条为《辞海》和《汉语大词典》收录，但其定义和词源完全是现代的。

43 佐藤道信（Dōshin Satō），*Modern Japanese Art and the Meiji State: The Politics of Beauty*, Hiroshi Nara 翻译（Los Angeles, CA: The Getty Research Institute, 2011 年），第 271 页。

44 许慎，《说文解字》，成书于公元 100—121 年（北京：中华书局，1963 年），12 篇上，第 248 页。在谈论“画”这个表意字的语境中，余定国（Cordell D.K. Yee）认为“《说文》在字源方面或许不是可靠的指南，但至少表达了汉朝人对字源的看法和误解。”这一观点不难令人接受。见 Cordell D.K. Yee, “Chinese Cartography among the Arts: Objectivity, Subjectivity, Representation,” 载于 *The History of Cartography* Vol. 2, Book 2, J. B. Harley、David Woodward 编 辑（Chicago, IL: Chicago University Press, 1994 年），第 139 页。

45 段玉裁（1735—1815 年）在其《说文解字注》（1808 年成书，并于 1815 年出版 [上海：上海古籍出版社，1981 年，12 篇上，第 589 页]）中指出，为了区分两者，当这个字用于第二种含义时，后来读音为 xián。在诸如《管子》和《吕氏春秋》之类的古典文献中，将“空”和“间”这两个汉字组合成一个词的情况很少，即使偶尔出现，也都表示“未占用”或“空闲”。

46 Burton Watson 翻译，*The Complete Works of Chuang Tzu*（New York: Columbia University Press, 1968 年），第 256 页。汉语文本，参见王先谦（1842—1917 年）《庄子集解》（北京：中华书局，1987 年），《杂篇 • 庚桑楚》，第 23，第 203 页。

47 马绛（John S. Major）等翻译、编辑，*The Huainanzi: A Guide to the Theory and Practice of Government in Early Han China*, by Liu An, *King of Huainan*（New York: Columbia University Press, 2009 年），第 415 页。在该书第三章中，马绛将“宇宙”简单地译为“时空”（space-time）（第 114 页以及同上页，注释 1），但“时空”一词不仅没能“很好地抓住其观念”，反而似乎更具误导性。其引用的汉语文本，可参见何宁《淮南子集释》（北京：中华书局，1998 年），卷 11，第 798 页。葛瑞汉（Angus C. Graham）在这个语境中对这些汉字的理解——“时间中持续的过程”与“空间中延展的事物”——相对来说更合乎情理，因为“时间”和“空间”这两个词在这里仅起间接作用，即暗示持续和延伸发生在我们现在泛泛所说的时间和空间中。参见 A.C. Graham, “Reflections and Replies,” 载于 *Chinese Texts and Philosophical Contexts: Essays Dedicated to Angus C Graham,* Henry Rosemont Jr. 编辑（La Salle, IL: Open Court, 1990 年），第 279 页。

48 何宁，《淮南子集释》，卷 6 第，469 页。马绛（Major, *Huainanzi*, 第 221 页）用以下方式翻译了这句话：“So the swallows and sparrows mocked them, saying that they were incapable of matching them in squabbling among the roof beams and rafters”。马绛接着说（同上，注 47）：“这个文本是文字游戏：‘宇宙’的字面意思是‘屋顶横梁和椽子（例如，谷仓的横梁和椽子，燕子和麻雀的自然栖息地），但它也提醒人们注意它更加惯用的‘宇宙’（即凤凰的栖息地）的含义”。他对文字游戏的识别确实颇有见地。然而我们很难将“宇宙”的含义视为“更加惯用的”——相反，“宇”和“宙”的“字面”意义一定更加惯用，否则高诱（活跃期公元三世纪初）和其他后来的评论者就无需在谈及“宇宙”时解释这两个字在那些语境下的衍生含义。

49 Richard E. Strassberg, *Inscribed Landscapes: Travel Writing from Imperial China*（Los Angeles, CA: University of California Press, 1994 年），第 138、143、163 页。重点符号是我添加的。参较欧阳修，《欧阳修全集》（北京：中华书局，2001 年），卷 39，第 576-577 页；白居易，《白居易集》（北京：中华书局，1979 年），卷 43，第

940-941 页；柳宗元，《柳宗元集》（北京：中华书局，1979 年），卷 29，第 765-766 页。

50 James Cahill, *Hills Beyond a River: Chinese Painting of the Yüan Dynasty, 1279-1368*（New York: Weatherhill, 1976 年），第 16 页。重点符号是我添加的。

51 郭若虚，《图画见闻志》，见于安澜编辑《画史丛书》（上海：上海人民美术出版社，1962 年），第 1 册，第 1 卷，第 6 页。

52 Alexander Coburn Soper, *Kuo Jo-Hsü's Experiences in Painting: An Eleventh Century History of Chinese Painting Together with the Chinese Text in Facsimile*(Washington, D.C.: American Council of Learned Societies, 1951 年），第 12 页。重点符号是我添加的。

53 Susan Bush、Hsio-yen Shih, *Early Chinese Texts on Painting*（Hong Kong: Hong Kong University Press, 2012 年），第 111 页。重点符号是我添加的。

54 Robert J. Maeda, "Chieh-hua: Ruled-line Painting in China," *Ars Orientalis* 10（1975 年）：第 123 页。重点符号是我添加的。

55 Caswell, "Lines of Communication," 第 201 页。重点符号是我添加的。

56 Duncan M. Campbell, "Zheng Yuanxun's 'A Personal Record of My Garden of Reflections'", *Studies in the History of Gardens & Designed Landscapes* 29, no. 4（2009 年）：第 280 页，注解 18。重点符号是我添加的。原文参见李斗著，《扬州画舫录》，汪北平、涂雨公点校（北京：中华书局，1960 年），城北录第六，第 145 页；李斗著，《扬州画舫录》（扬州：江苏广陵古籍刻印社，1984 年）城北录第六，第 139 页（坎贝尔参考了此书）。

57 李惠仪（Wai-Yee Li），"Gardens and Illusions from Late Ming to Early Qing", *Harvard Journal of Asiatic Studies* 72, no. 2（2012 年）：第 328 页。重点符号是我添加的。原文载于祁彪佳，《祁彪佳集》（北京：中华书局，1960 年），卷 7，第 167 页。

58 Wen Fong, *Beyond Respresentation: Chinese Painting and Calligraphy 8th-14th Century*（New York: The Metropolitan Museum of Art, 1992 年），第 85 页。关于研究的段落，参见郭熙、郭思，《林泉高致》，载于黄宾虹、邓实编辑《美术丛书》第 17 册（杭州：浙江人民美术出版社，2013 年），第 17 页。

59 Mitrović, "Leon Battista Alberti," 第 425、429 页。

60 Pierre Ryckmans, "The Chinese Attitude Towards the Past," *Papers on Far Eastern History* 39（1989 年 3 月）：第 8 页。

61 孙诒让（1848—1908 年），《周礼正义》（北京：中华书局，1987 年），卷 83，第 3423-3430 页。我的英文翻译基于 Paul Wheatley 的翻译，*The Pivot of the Four Quarters: A Preliminary Enquiry into the Origins and Character of the Ancient Chinese City*（Edinburgh: Edinburgh University Press, 1971 年），第 411 页。法文翻译，参见 Edouard Biot, *Le Tcheou-li ou Rites des Tcheou,* tome II（Paris: Imprimerie nationale, 1851 年），第 556 页。《考工记》是规范性、指示性的文本，而不是历史性文本，于汉武帝（公元前 141—前 87 年在位）时期根据战国后期的文本制定。

62 关于园林文献中地图方式与游历方式运用的讨论，参见 Craig Clunas, *Fruitful Sites: Garden Culture in Ming Dynasty China* (London: Reaktion Books, 1996 年)，第 140-144 页；许亦农（Yinong Xu），"Boundaries, Centres, and Peripheries in Chinese Gardens—A Case of Suzhou in the Eleventh Century," *Studies in the History of Gardens and Designed Landscape* 24, no. 1（2004 年 3 月）：第 31-32 页。

63　朱长文，《乐圃记》，载于范成大（1126—1193年），《吴郡志》（南京：江苏古籍出版社，1986年），卷14，第194-196页。关于乐圃布局的文字描述的分析，参见许亦农，“Boundaries,”第30-33页。

64　还值得注意的是，在袁黄（1533—1606年）撰写的《坐隐先生环翠堂记》中这四个区域之间文字上的关系与手卷中的画面上的毗邻性十分不同：“环翠堂”的前庭和“兰亭遗圣”的描述在文本中出现得很早，与落笔于文本约三分之二位置的另两处的描述距离相当远；而后出现的这两处区域的先后顺序也与卷轴中的刻画顺序相反。见袁黄，《坐隐先生订谱》中《坐隐先生环翠堂记》，汪廷纳（十六世纪末至十七世纪）编辑（环翠堂版本，约1609年），匏部，页35a-页40a。此文本的翻译，见 Alison Hardie, “Record of the Hall Surrounded by Jade of Master Sitting-in-Reclusion,” 载于 *The Dumbarton Oaks Anthology of Chinese Garden Literature*, Alison Hardie、Duncan M. Campbell 编辑（Washington D.C.: Dumbarton Oaks, 2020年），第381-392页。

65　Massimo Scolari, *Oblique Drawing: A History of Anti-Perspective*（Cambridge, MA: The MIT Press, 2012年），第348页。

66　John Hay, “Chinese Space in Chinese Painting,” 载于 *Recovering the Orient: Artists, Scholars, Appropriations*, Andrew Gerstle、Anthony Milner 编辑（Amsterdam: Harwood Academic Publishers, 1994年），第151页。

CONTRIBUTORS
作者简介

CHEN Wei
陈　薇

Professor CHEN Wei teaches in the School of Architecture, Southeast University, and is Director of the Institute of Architectural History and Theory. She is an academic leader in this discipline, and has published extensively on Chinese garden history.

陈薇教授，执教于东南大学建筑学院，并任建筑历史与理论研究所所长，是该学科学术带头人。在中国园林史方面著述颇丰。

GE Ming
葛　明

Professor GE Ming, PhD, is Associate Dean International of the School of Architecture, Southeast University. His main research area is architectural and garden design, as well as architectural theory, and his design work includes Wei yuan and Chun yuan.

葛明教授，博士，东南大学建筑学院副院长，主要研究方向为建筑与园林设计、建筑理论，主要设计作品有微园、春园等。

Antoine GOURNAY
顾乃安

Professor Antoine GOURNAY teaches Far Eastern Art and Archaeology at Sorbonne Université, Paris. His research field is the history of Chinese architecture and gardens. Related fields of interest include landscape painting in China, the production and use of ceramics and furniture, and the archaeology of religious cults in East Asia.

顾乃安教授，在巴黎索邦大学远东艺术和考古学任教，研究领域为中国建筑史与园林史，其他相关研究领域包括中国山水画、陶瓷和家具的制造与使用，以及东亚宗教崇拜考古学。

GU Kai
顾　凯

Associate Professor GU Kai, PhD in Architectural History and Theory from Southeast University, where he is now teaching in the Department of Landscape Architecture, School of Architecture. His main research area is history, theory and heritage conservation of landscape architecture.

顾凯副教授，东南大学建筑历史与理论博士，现任教于东南大学建筑学院景观学系，主要研究方向为风景园林历史、理论与遗产保护。

Alison HARDIE
夏丽森

Dr Alison HARDIE was until 2015 a Senior Lecturer in Chinese Studies at the University of Leeds, UK, where she is now an Honorary Research Fellow. Her main research interest is in the social and cultural history of late imperial China. She has translated and written extensively on Chinese garden history.

夏丽森博士，迄 2015 年担任英国利兹大学中国研究高级讲师，目前为该大学名誉研究员。主要研究兴趣为中华帝国晚期的社会与文化史，在中国园林史方面翻译和著述颇丰。

Minna TÖRMÄ
米娜·托玛

Dr Minna TÖRMÄ studied art history and theatre history at University of Helsinki and received her PhD with a dissertation on Northern Song landscape painting ("Landscape Experience as Visual Narrative") in 2002. She currently works as Lecturer in History of Art at the University of Glasgow. She is also Adjunct Professor of Art History at University of Helsinki. Her latest publication is *Nordic Private Collection of Chinese Objects* (Routledge 2020).

米娜·托玛博士曾在赫尔辛基大学研习艺术史与戏剧史；2002 年获得博士学位，研究北宋山水画（"作为视觉叙事的景观体验"）；现任格拉斯哥大学艺术史讲师，赫尔辛基大学艺术史兼职教授。最新著作为 2020 年劳特里奇出版的《北欧个人收藏的中国物件》。

XU Yinong
许亦农

Professor XU Yinong, PhD in Architectural and Urban History from the University of Edinburgh, now Director of Caxton House—China Research and Engagement, London South Bank University. His area of research interest centres on Chinese architectural history, urban history, and garden history from cross-cultural and interdisciplinary perspectives.

许亦农教授，爱丁堡大学博士，现任伦敦南岸大学卡克斯顿中国研究与交流学院院长，主要研究领域：跨文化、跨学科视野下的中国建筑史、城市史和园林史。

图书在版编目（CIP）数据

胸中丘壑：中国园林与山水画 = Hills and Ravines in the Heart: Chinese Gardens and Landscape Painting : 汉、英 / 许亦农，葛明，顾凯编. -- 南京：东南大学出版社，2023.11
ISBN 978-7-5766-0239-5

Ⅰ. ①胸… Ⅱ. ①许… ②葛… ③顾… Ⅲ. ①园林艺术 - 关系 - 山水画 - 绘画研究 - 中国 - 汉、英 Ⅳ. ① TU986.62 ② J212.26

中国版本图书馆 CIP 数据核字（2022）第 171873 号

责任编辑：戴　丽　魏晓平
责任校对：张万莹
封面设计：刘筱丹
责任印制：周荣虎

胸中丘壑：中国园林与山水画
Xiong Zhong Qiuhe: Zhongguo Yuanlin Yu Shanshui Hua

出版发行：东南大学出版社
出 版 人：白云飞
社　　址：南京市四牌楼 2 号
网　　址：http://www.seupress.com
邮　　箱：press@seupress.com
邮　　编：210096
电　　话：025-83793330
经　　销：全国各地新华书店
印　　刷：上海雅昌艺术印刷有限公司
开　　本：787 mm × 1 092 mm　1/16
印　　张：22
字　　数：498 千
版　　次：2023 年 11 月第 1 版
印　　次：2023 年 11 月第 1 次印刷
书　　号：ISBN 978-7-5766-0239-5
定　　价：198.00 元